CONSERVING SOUTHERN LONGLEAF

environmental
history
and the
american
south

CONSERVING SOUTHERN LONGLEAF

Herbert Stoddard and the Rise of Ecological Land Management

ALBERT G. WAY

THE UNIVERSITY OF GEORGIA PRESS

ATHENS AND LONDON

Portions of chapter 3 appeared previously as "Burned to Be Wild: Herbert Stoddard and the Roots of Ecological Conservation in the Southern Longleaf Pine Forest," *Environmental History* 11, no. 3 (July 2006): 500–526. Reprinted by permission of Oxford University Press.

Athens, Georgia 30602
www.ugapress.org

Set in New Caledonia by Bookcomp, Inc.

Printed digitally

Library of Congress Cataloging-in-Publication Data
Way, Albert G.
Conserving southern longleaf : Herbert Stoddard and the rise of ecological land management / Albert G. Way.
p. cm. — (Environmental history and the American South)
Includes bibliographical references and index.
ISBN-13: 978-0-8203-3466-0 (hardcover : alk. paper)
ISBN-10: 0-8203-3466-9 (hardcover : alk. paper)
ISBN-13: 978-0-8203-4017-3 (pbk. : alk. paper)
ISBN-10: 0-8203-4017-0 (pbk. : alk. paper)
1. Longleaf pine—Conservation—Red Hills (Fla. and Ga.) 2. Forest management—Red Hills (Fla. and Ga.) 3. Stoddard, Herbert L. I. Title.
SD397.P59W39 2011
634.9'2—dc22 2011010450

British Library Cataloging-in-Publication Data available

Publication of this book was made possible, in part,
by a generous gift from Alan Rothschild
for the Fort Trustee Fund.

For My Parents

CONTENTS

ILLUSTRATIONS

FOREWORD

Anybody who has read widely in American environmental history has come across Herbert Stoddard's name. Stoddard has made cameo appearances in some of the most important books in the field. In her pioneering study of Aldo Leopold's evolving thought on predator-prey relations, *Thinking like a Mountain*, Susan Flader noted in passing that "unquestionably the most significant research in early game management was by Herbert L. Stoddard." In *Fire in America*, Stephen Pyne argued that it was Stoddard who "gave scientific credibility to the subject of wildlife management through fire." In *Saving America's Wildlife*, his influential history of American wildlife conservation, Thomas Dunlap noted that Stoddard's 1931 study, *The Bobwhite Quail: Its Habitats, Preservation, and Increase*, "set the example and stimulated a more rigorous science of game management." And Curt Meine, in his still-definitive biography of Aldo Leopold, noted that Leopold "personally held that Stoddard was the true pioneer" of the practice of game management. "While Leopold was evolving an abstract framework for the science," Meine insisted, "Stoddard was providing its first concrete example."[1] But while we have long known about Stoddard's importance—as a founding practitioner of game or wildlife management and as a seminal figure in the scientific reconsideration of fire as an ecological agent—until now we have lacked a careful study of Herbert Stoddard's achievements and influence as an ecological thinker and practical land manager. Giving Herbert Stoddard his due—taking him from the shadows of environmental historiography and placing him deservedly in the sun—is one of the most important achievements of Bert Way's wonderful new book *Conserving Southern Longleaf*.

But *Conserving Southern Longleaf* is much more than an environmental biography of Herbert Stoddard; Way's study is also an innovative place-based history of Stoddard's achievements and influence. This is a book about where conservation science comes from, how it springs from particular landscapes and communities. Stoddard's story is inextricable from the natural and human history of the Red Hills, a remarkable region of undulating longleaf pine woodlands that stretches between Thomasville, Georgia, and Tallahassee,

Florida. Stoddard's life and work "took place" in the Red Hills in the richest sense of that phrase; his achievements, until now so vaguely understood, were the products of a lifelong dialogue with this remarkable natural community that was his working environment. And the Red Hills landscape into which Herbert Stoddard moved in the early 1920s was at the center of an imperiled biome—the longleaf pine woodlands that once cloaked seventy to ninety million acres of the southern coastal plain from Virginia to Texas and helped to define the natural South. Historically one of the North American continent's most dominant and biodiverse natural communities, longleaf pine woodlands now exist only in remnants, and high-quality examples are frighteningly rare. Herbert Stoddard not only gave us a critical start at understanding and appreciating this endangered natural system, he also gave birth to a land management approach that was well ahead of its time. Indeed, many of the best remnants of longleaf pine woodlands are still intact only because of Herbert Stoddard and his larger managerial influence. *Conserving Southern Longleaf*, then, is also an important environmental history of a unique pocket of the American South—and of the larger longleaf bioregion in which it is nestled.

But wait, there's more: in telling us the story of Stoddard and his place, Way provides us with a surprisingly revisionist history of American conservation. To a certain degree, Way's study follows a plotline with which environmental historians are familiar. *Conserving Southern Longleaf* is a story about how creative and dedicated conservationists—Stoddard and his Red Hills cohort—came to understand and appreciate the ecological and aesthetic value of a particular natural system and then worked for its protection. But the details of this conservation story will be, to many environmental historians, provocative and even jarring. This is not a story of a heroic federal conservation intervention and the protection of public lands; while Stoddard did come to the Red Hills under federal auspices, and while his work intersected with and influenced federal conservation policy throughout his half-century in the Red Hills, he did most of his work on private lands, much of it as a private citizen. There is no national park at the end of this story; the landscape legacies of Herbert Stoddard's work in the Red Hills are more complicated than that. In that sense, *Conserving Southern Longleaf* provides a particularly southern environmental narrative, one focused on private-lands conservation.

But perhaps the greatest achievement of *Conserving Southern Longleaf* is how successfully and meaningfully Way is able to detach conservation history from the wilderness ideal that has long preoccupied environmental historians. Even in their most "natural" expressions, the longleaf pine woodlands with which Stoddard worked were, at base, Holocene creations that developed in concert with human land use. As the products of frequent fire, these longleaf woodlands were shaped by intertwined natural and human histories. More importantly, when Stoddard came to the Red Hills in the 1920s, he found a landscape with a more recent disturbance regime of small-scale agricultural clearances created by tenant farming. Although he was working for the Bureau of Biological Survey (which later became the Fish and Wildlife Service), Stoddard came to the area at the behest of wealthy landowners who had bought up old plantations as quail hunting preserves and who were struggling to maintain quail numbers. After carefully studying the bobwhite quail and its habits, Stoddard made several broad conclusions. First and most memorably, Stoddard insisted that the region's longleaf woodlands needed to be burned regularly or they would grow up into a "rough" of successional hardwoods that would diminish quail numbers and, he later realized, the remarkable diversity of the longleaf understory. Fire was also critical for longleaf reproduction. At a moment when natural fires no longer proved capable of maintaining the historically open longleaf woodlands of the region, human management would have to step in. This was not a landscape in which meaningful ecological protection could happen by leaving nature alone.

Stoddard's suggestion that managers burn longleaf woodlands was only heretical among conservation experts, who were then hard at work trying to eradicate southern woodsburning practices that they characterized as pathological. Local people, black and white, had been burning the woods for a long time, and Stoddard was quick to see the wisdom of their habits. Moreover, Stoddard quickly realized that bobwhite quail—a disturbance species that thrived on early successional plant communities—also relied on the recent-historical land-use regimes and practices of the Red Hills. In other words, the longleaf woodlands of the Red Hills were great quail habitat not because of their pristine state, if there even was such a thing, but because tenant farming had provided quail with the food and cover they needed to thrive. These were working and peopled landscapes of conservation, rich in the social and racial conflicts that defined the modern South. To maintain and even build quail populations, Stoddard realized, such disturbances would have to continue in one form or another. Indeed, even as his management goals expanded to include goals beyond quail maximization, Stoddard always insisted that the best way to maintain an ecologically functional and aesthetically pleasing longleaf landscape was to continually disturb it—wisely and humbly—with fire, highly selective timber cutting, and other sorts of small-scale clearing. Leaving these woodlands to be governed by natural forces alone was the last thing that an intelligent conservationist wanted to do.

To come to know Herbert Stoddard, then, is to come to know a whole new culture of conservation, one particular to the Red Hills, to the natural history of the larger longleaf region, and to the human history of southern agricultural landscapes and practices. Moreover, it is to come to know a conservation tradition that is a critical model for present and future forms of ecological land management. While the work of Herbert Stoddard, though recognized in passing, may have been hard to appreciate in an earlier generation of environmental history scholarship, Stoddard's story is remarkably salient today, in an age of postwilderness conservation.

Paul S. Sutter

1. Susan Flader, *Thinking like a Mountain: Aldo Leopold and the Evolution of an Ecological Attitude toward Deer, Wolves, and Forests* (Madison: University of Wisconsin Press, 1974), 24; Stephen Pyne, *Fire in America: A Cultural History of Wildland and Rural Fire* (Princeton, N.J.: Princeton University Press, 1982), 154; Thomas Dunlap, *Saving America's Wildlife: Ecology and the American Mind, 1850–1990* (Princeton, N.J.: Princeton University Press, 1988), 71; and Curt Meine, *Aldo Leopold: His Life and Work* (Madison: University of Wisconsin Press, 1988), 264.

ACKNOWLEDGMENTS

I started thinking about this book while conducting background reading for another project, an oral history with Herbert Stoddard's protégé and friend Leon Neel. I began with Stoddard's memoir, a wonderful reflection on a life well lived but not, I thought, quite the stuff of a dissertation. I then turned to Stoddard's *The Bobwhite Quail*, a staggering piece of work that located the South as a major center for modern American conservation—a perspective I was beginning to think simply did not exist. The cultural landscapes of the Red Hills leaped off the pages in such a way that immediately set me to thinking about how this place might complicate and, in many ways clarify, the history of American conservation. I still wasn't convinced, however, until I, along with my graduate advisor Paul Sutter, visited Leon Neel. Leon led us through some of the most outstanding longleaf woodlands in the Southeast, and that sealed the deal.

That first visit with Leon has led to several fulfilling years, both professionally and personally. Leon, Paul, and I converted Leon's oral histories into a memoir and published it in 2010 as *The Art of Managing Longleaf: A Personal History of the Stoddard-Neel Approach*. And now this book is finally seeing print as well. More important, though, I've had the chance to become friends with Leon and his wife, Julie. They have welcomed me into their home, fed me, and given me a place to lay my head; they have been remarkable guides in the field; and they have shared stories with me about Herbert Stoddard, as well as their own lives. They have also read drafts of this book and offered words of both encouragement and correction. This book is much better because of their involvement with the project.

Paul Sutter has been with this project from the beginning, and his guidance as mentor, editor, and friend has been exceptional. He read numerous drafts and always delivered lengthy feedback that tightened up my prose and analysis and usually led me to see things in a different light. I've been fortunate to have him on my side.

The history department at the University of Georgia was an outstanding intellectual home for seven years. Jim Cobb, John Inscoe, Allan Kulikoff, Shane Hamilton, and many other faculty members offered wise counsel time and again on not only this manuscript but also my career. Others at UGA who influenced my thinking, and my drinking, include Judkin Browning, Amy Crowson, Frank Forts, Ed Hatfield, John Hayes, Chris Huff, Christopher Lawton, Robby Luckett, Alex Macaulay, Jason Manthorne, Ichiro Miyata, Steve Nash, Justin Nystrom, Tom Okie, Lesley-Anne Reed, Levi Van Sant, Hayden Smith, Bruce Stewart, and Keira Williams. Special thanks to Jim Giesen and Chris Manganiello for reading much of this work and for the many serious discussions we've had over the years about southern environmental history.

The Institute for Southern Studies at the University of South Carolina was an ideal place to revise this manuscript for publication. Walter Egdar, Bob Ellis, Bob Brinkmeyer, Thorne Compton, Tara Powell, Mindi Spencer, and Walter Liniger were wonderful hallmates for two years. I am grateful too to the Watson-Brown Foundation for funding such a unique postdoctoral fellowship. In the history department at USC, Emily Brock, Don Doyle, and Ann Johnson were welcoming, and Tom Lekan's generosity of time and spirit always left me overcome with appreciation.

The Joseph W. Jones Center for Ecological Research at Ichauway has been a sort of part-time institutional home for me over the past several years. More than that, it is a spectacular place to find inspiration. Lindsay Boring offered unwavering support from beginning to end and has been generous enough to lend me the title of "Visiting Scientist" from time to time, a line on my CV that has raised some interesting questions during job interviews. Kevin McIntyre was my guide to all things southwest Georgia. He set me up for a six-month residency at Ichauway during the early stages of this project and gave first-rate advice at key moments along the way. Steve Jack, Bob Mitchell, Kay Kirkman, Lora Smith, Steve Golladay, Mike Smith, Woody Hicks, Jimmy Atkinson, and Mark Melvin were always patient with my lay questions about science and land management in the longleaf pine region. Thanks also to Sue Hilliard and Denise McWhorter for their skills of organization. One of the great things about environmental history is getting out into the field and learning about particular landscapes, and the staff, graduate students, and residents of Ichauway have been exceptional guides through the region.

Several scholars have been important influences and have helped me to make the transition into professional life. Jack Temple Kirby's work was an early touchstone for me, only surpassed in quality by his company. I did not know him for very long, but I'm thankful for the moments we shared. Tim Silver commented on an early conference paper I delivered on Stoddard and said some of my ideas might "twist our southern wigs," a comment that made me nervous at the time but that I have since savored. Mart Stewart continues to be one of the guiding voices of southern environmental history, and his encouragement has meant the world to me. I would also like to thank Claire Strom and Mark Finlay for helping in a number of ways and Shepard Krech for recounting his wonderful memories of the Red Hills. For commenting on all or parts of the manuscript, I also thank Mark Barrow, Fritz Davis, Mark Hersey, Rob Kohler, James Pritchard, and many other participants at conferences and workshops where I have presented this work.

As fun as environmental history is in the field, the archive is still our place of business. First, I thank those at Tall Timbers Research Station who gave me access to its remarkable manuscript collection. Not only is its archive an important resource for environmental history, its archivist, Juanita Whiddon, made using the archive a pleasure. She opened up the doors to the place and was always available to answer questions and direct me to where I needed to be. Also, Jim Cox, Lane Green, Ron Masters, and Kevin Robertson took an interest in the project and gave encouragement when called on. I also thank Cheryl Oakes, Steven Anderson, and Jamie Lewis at the Forest History Society, Ann Harrison and Ephraim Rotter at the Thomas County Historical Society, and Liz Cox at Ichauway for fielding requests from near and far. Archivists at the University of Wisconsin, Madison, the Field Museum in Chicago, and the Milwaukee Public Museum made my swing through the Midwest productive and rewarding. Also, thanks to Curt Meine for answering many questions about Aldo Leopold, his papers, and the field of conservation biology in general.

The staff at the University of Georgia Press has made the publishing process an enjoyable one. Thanks to Nicole Mitchell for her support of environmental topics, Derek Krissoff for his excellent editorial guidance, and Jon Davies and Sue Breckinridge for their attention to detail.

As important as all of the above have been, my family has been my most important influence. My brothers, Ramsey and Hatcher, and my sister, Elizabeth, patiently withstood their annoying little brother until I could figure things out for myself. They and their families have been incredibly supportive. And my parents, Mary Lynde and Bub, have always been inspirations to me.

Finally, my deepest gratitude goes to Ivy Holliman Way. She endured my trips to south Georgia and my frequent talk about longleaf pines with good humor; and she read much of the manuscript and helped me work through some of its most prickly problems. But more than that, she has improved my life in ways immeasurable, for which she has not only my thanks but also my complete devotion.

CONSERVING SOUTHERN LONGLEAF

INTRODUCTION

Just outside of the city limits of Thomasville, Georgia, runs a public dirt road called Pine Tree Boulevard. It was once a perimeter road, but in recent years the city moved parts of it due to various zoning or planning schemes. There remains, however, one section about a mile in length that is surrounded by the most spectacular longleaf pine woodland one will ever see. The forest is part of Greenwood Plantation, a former antebellum plantation purchased by wealthy northerners a few years after the Civil War. The big house and surrounding structures are impressive and have an important place in this story, but it's the forest that is most stunning.

Like anyone who drives the thoroughfares and back roads of the southeastern coastal plain, I am familiar with what is classed as the region's "forest land." And it does not look anything like Greenwood. The oldest longleaf pines on Greenwood are more than four hundred years old and stand with their gnarly flat tops at over one hundred feet tall. Younger pines of various ages are scattered in piecemeal groups throughout the midstory canopy—nothing like the uniform rows of planted pines that occupy so much of the southern landscape. The forest is open, and sunlight streams down in waves through the canopy, easily reaching the lush grasses and legumes of the understory. Longleaf pine woodlands have a claim as the most biologically diverse environment in temperate North America, and Greenwood is one of the finest remaining examples. When I first encountered Greenwood, a fairly straightforward question came to mind: how did this piece of land, and a few others with similar appearance and provenance, escape the myriad economic and social forces that transformed the rest of the southern countryside throughout the long twentieth century? The quick answer is that shortly after the Civil War, the nation's wealthiest industrialists sought retreat in this area surrounding Thomasville, Georgia, and Tallahassee, Florida, an area known as the Red Hills for its gently rolling terrain and red clay soils. They purchased large swaths of former plantation land, kept it free of human manipulation, and thus allowed the forests to grow unmolested. The complete answer, however, is much more complicated. These lands never lay outside the realm of market economics, and their ecosystems were certainly never free from human action. On the contrary, these longleaf forests remain as they are *because* of human manipulation. The word *manipulation* usually indicates devious intent, often rightfully so. But at least one definition is more ambiguous, and in this case more appropriate: "skillful or artful management."[1] When applied to human relationships, and even some human-environmental relations, a certain Machiavellian quality lingers; but in a place like Greenwood, skillful and artful management implies a knowledgeable mind, a discerning eye, and a caring hand.

Today in the longleaf woodlands of places like Greenwood, those who work the land usually seek to maintain or restore native biological diversity. But that was not always the case. Diversity in the Red Hills was once a by-product of a management with different designs—designs that had productivity and profit in mind. When wealthy northerners found a winter's stay in the Red Hills to be a healthful respite from their region's growing industrial centers, planters and former slaves were busy negotiating a system of tenantry and sharecropping that would continue to dominate the Red Hills countryside well into the twentieth century. Into the 1880s and 1890s the exhilarating hunt of the bobwhite quail amid stately longleaf pine woodlands brought visitors back again and again, and they took advantage of the precarious position of planters to eventually purchase about three hundred thousand acres of plantation land, largely leaving tenant arrangements just as they were. The shift toward managing this land for biological diversity, then, emerged as a hybrid system rooted in the productive processes of southern agriculture and the aesthetic ideologies of northern industrialists. But it also emerged as field scientists began to build a deeper understanding of the region's environmental processes.

This book charts several local, regional, and national trends in science and conservation, and it offers an alternative path toward what may be called an ecological consensus in American environmental thought. This consensus refers to the ideas of a loosely organized coalition of scientists, nature lovers, and ordinary citizens who began to see the world through an ecological lens in the early twentieth century and who rose out of the World War II–era "thinking like a mountain," to use Aldo Leopold's evocative language. The rise of this modern environmental consciousness began with Progressive-era concerns for dwindling resources and the concomitant growth in knowledge about how natural environments worked. Government experts helped to create administrative state structures to efficiently protect and control land and water resources, while academic biologists created the organizational framework necessary to discover, test, and disseminate scientific knowledge about nature.[2] As these state and academic entities gained capacity, the American public joined the fray as well, growing increasingly interested in nature, and concerned about losing the nation's former natural bounty. The public also worried over their ability to determine control of local resources. Indeed, many communities resisted state authority in local environments, as well as the efforts of scientific experts to direct local land-use practices. In some areas, state conservation projects facilitated the movement of people from the countryside, disrupting longtime agrarian traditions and turning lives upside down.[3] In other areas, on the other hand, communities embraced natural-resource conservation through state making and actually directed much of the project.[4]

Within this project of government conservation, there developed a concern for nature on its own terms. Scientists and naturalists during the interwar years, in particular, became anxious about the environmental effects of industrialization and modernization, as well as what these modes of production and consumption meant for the ways Americans experienced nature. In constructing a scientific base to explain nature, biologists attempted to lend credibility to a moral concern for nature in decline.[5] In the midst of this knowledge making came a tremendous expansion of governmental capacity to control the nation's natural resources. New Deal programs such as the Agricultural Adjustment Administration told farmers what and how much to plant, the Soil Conservation Service suggested to them how to plant, the Resettlement Administration removed people from worn out lands, and the Civilian Conservation Corps sent men into the countryside to reclaim those lands. While most economists and land planners applauded these efforts, some scientists and naturalists grew concerned that government administrators were making the wrong choices about managing nature. They wondered if the attempts of state agencies to protect nature were actually doing it harm.[6]

This composite narrative of the rise of American environmental thought is accurate enough, but the ways that scientists and laypersons came to know nature in an effort to save it took many different forms—forms that are often obscured by environmental history's focus on wilderness, state conservation, and public lands. This book explores the development of ecologically based conservation and science in a different context—the thoroughly worked fields and forests of the American South's coastal plain, where longleaf pine woodlands constituted the dominant historical environment. On the Red

Hills hunting preserves, in particular, a variety of actors devised and applied a unique form of conservation, and one of my goals here is to explain the complex processes through which a handful of scientists questioned, tested, and finally codified local environmental knowledge into a kind of hybrid science. In the years covered here, roughly 1880–1960, the Red Hills was a working landscape that produced a variety of goods and supported a large number of people. At the same time it was a conservation landscape that harbored some of the most ecologically sound examples of a fading environment, as well as a laboratory where a great deal of scientific knowledge about the longleaf pine biome came to light.[7]

The central figure here is Herbert L. Stoddard, an ornithologist and wildlife biologist who came to the Red Hills in 1924 as an agent of the U.S. Biological Survey to examine the life history and preferred habitat of the bobwhite quail and to develop a system of management to reverse quail population declines. He remained until his death in 1970, all the while developing a land management program focused on maintaining and restoring ecological integrity and simultaneously engaging in the processes of production. His scientific work on fire ecology, predator-prey relations, wildlife management in agricultural systems, and ecological forestry all mirrored the national move toward a more biocentric vision of nature. But like many who developed a moral concern for nonhuman nature during the interwar years, Stoddard still considered the environments he worked in and on to be a bundle of natural resources, and his constituencies continued to plow the earth, cut the forests, and shoot, trap, and snare the animals. One argument of this study, then, is that conservation of the land and work on the land were not always mutually exclusive, nor were they always in opposition. The biocentric turn in twentieth-century science and society was rooted as much in the processes of production that it sought to moderate as it was in a new philosophical ideal about nature's inherent value.[8]

Herbert Stoddard's story is remarkable, and on one level this book serves as his intellectual biography. But it works on deeper levels as well. It explores the local, regional, and national implications of his work in a natural system that once dominated the southern coastal plain, and it uses his life and work as a window into the social and environmental realities that created a conservation landscape in the Red Hills. I also examine the broader conservation and scientific tradition from which Stoddard came and the important contributions of his work to the growth of ecological conservation in the United States and beyond. Stoddard came of professional age alongside some of the most accomplished naturalists and scientists of their day—including his close friend Aldo Leopold, perhaps the most important environmental thinker of the twentieth century—yet he also worked alongside local people who shaped their surrounding environments with purpose. He learned from and influenced both groups equally. Through a focused study of Stoddard and his social, professional, and environmental surroundings, we see a form of conservation that was not imposed as a high-modernist plan of the administrative state or created to meet the economic demands of an industrial ideal. Nor did conservation in the coastal plain develop solely through grassroots organization from the bottom up. Conservation and science in the Red Hills fell somewhere in between. It emerged in the interwar years as a union of the agrarian traditions of local southerners, the aesthetic ideologies of northern industrialists, and the insights and methodologies of the burgeoning science of ecology.

Conservation in the Red Hills also came to depend on the contours of the southern racial, social, and economic hierarchy in its organization and implementation. I began this journey on Pine Tree Boulevard, a public thoroughfare, because this is the only vantage point from which the majority of Thomasville's residents have ever caught a glimpse of Greenwood's forest. This was and is private land, and anyone raised in the modern South knows better than to cross a posted property line without permission. They also know that any piece of land that looks like this will be fiercely guarded. And it was, even in representation. Unlike public conservation landscapes, the expanding boundaries of southern hunting preserves in the Red Hills and other locales in the early twentieth century were not visible on public maps. The maps that did appear circulated only among the landowners themselves, and they came to represent prestige, exclusivity, and control. And if the control of people is often predicated on the control of land, the Red Hills presents a clear case of how a group of very powerful people secured control of the countryside for the use of natural resources and in the process shaped the livelihoods of the people remaining.[9] While most locals rarely set foot on these places, tenant farmers continued to work the land in much the same fashion, and under the same economic arrangements, as they had since Reconstruction. The plantation South's rural labor system, and its "varied but unpatterned blend of illiteracy, law, contracts, and violence," in the words of historian Pete Daniel, proved a comfortable organizational framework for conservation in the Red Hills.[10] It allowed for the close management of both human and natural resources.

The social, economic, and environmental realities of the South, then, shaped this conservation regime in important ways. Many human activities not normally associated with ecological health took place within these landscapes: tenant agriculture, staple crop production, racial division, private landownership, wealth, poverty—it was all there within the borders of these reserves of biological diversity. The South was an agrarian land, and as Mart Stewart has argued, the region never developed "an indigenous notion of 'wilderness' as unoccupied or relatively undisturbed nature."[11] Southern life in the Red Hills did, however, foster an indigenous form of ecological conservation, one that was agrarian in its sensibilities. When Herbert Stoddard began to couch the management decisions of landowners in ecological terms, he did so from within the context of southern agriculture, and his work in the Red Hills provides a prime case of how new environmental concerns can manifest themselves in place-specific ways.[12]

Stoddard was well connected to the national conservation establishment, but many of the practices he advanced came directly from the economic and cultural activities of southern rural life. The use of fire to maintain the longleaf woodlands was the most obvious ecological use of a cultural practice. Beyond fire, though, Stoddard sought to mimic many of the landscape components of tenantry in his land management. From the patchy agricultural fields to the night hunting of small mammals, he found many agricultural practices and subsistence strategies to be effective conservation measures in the longleaf pine region. Moreover, the social relations of tenantry made for conservation's smooth implementation. Unlike so many other conservation regimes that facilitated the removal of local people, conservation in the Red Hills actually capitalized on local social and economic arrangements to maintain the region's ecological integrity.[13] Without people living and working in this landscape, the region would not only have looked vastly different but the processes of learning about the region's ecology would have been different as well.

The type of conservation science Stoddard practiced in the Red Hills was also profoundly influenced by the scientific and environmental context of Progressive-era North America, and it came into its own during the interwar years. Stoddard developed his environmental sensibilities during a time that historian Robert Kohler calls the "age of the survey," the time between 1880 and 1930 when "scientists became fully aware of the world's biodiversity."[14] Essential to the surveyor's conception of the natural world was what Kohler calls modernizing America's "inner frontiers." This idea is an important one. The inner frontiers were the relatively uninhabited yet easily accessible spaces where natural processes carried on amid modernity. As Kohler explains, during the decades around the turn of the twentieth century, "densely inhabited and wild areas were jumbled together. Areas of relatively undisturbed nature, with much of its original flora and fauna intact (except for large game animals and predators), were accessible to people who lived in towns and cities, with their cultural and educational institutions. It was this combination of wildness and accessibility that defined the inner frontiers."[15] The age of the survey was about collecting and cataloging the pieces of nature in these inner frontiers, activities that set the stage for a new scientific understanding of the natural world in the age of ecology. Indeed, Herbert Stoddard's career bridged these two scientific ages; he was a transitional figure between a taxonomic and ecological understanding of nature. He trained as a taxonomic surveyor, but he became an ecological practitioner.[16]

The ecological and historical context of the Red Hills frames this story. Historically, the Red Hills was a small part of a ninety-million acre fire-dependent longleaf pine ecoregion that stretched across the southern coastal plain from southeastern Virginia to east Texas. It was one of North America's largest precontact natural communities, and today it barely hangs on as a viable ecological entity. As in so many other distinctive natural systems, successive waves of timber, forestry, agricultural, and other land-shaping interests have transformed the longleaf pine region over the past two centuries. By the time Stoddard came along in 1924, the Red Hills was one of the few places on the coastal plain where the longleaf pine woodlands and savannahs remained intact, thus giving it the critical qualities of an inner frontier. Stoddard's work in this inner frontier helped to shape what we currently know about the ecology of the longleaf system, and the following chapters reveal in layers how he and others produced that knowledge in the first half of the twentieth century. But as tempting as it is to cling to a revelatory narrative, not giving anything away until my actors learn it for themselves, it is difficult to understand the scope of their project—and the stakes involved—without some foregrounding of the longleaf system's historical ecology. So I travel back in time a hundred million years or so.[17]

As important as fire would become in the development of the coastal plain and its vegetative communities, water actually had more to do with its early creation. From the late Cretaceous Period until the last glacial period about eighteen thousand years ago, the seas that we know as the Atlantic Ocean and the Gulf of Mexico swamped the southern coastal plain dozens of times. Water made its furthest inland mark at the fall line, the meandering ridge

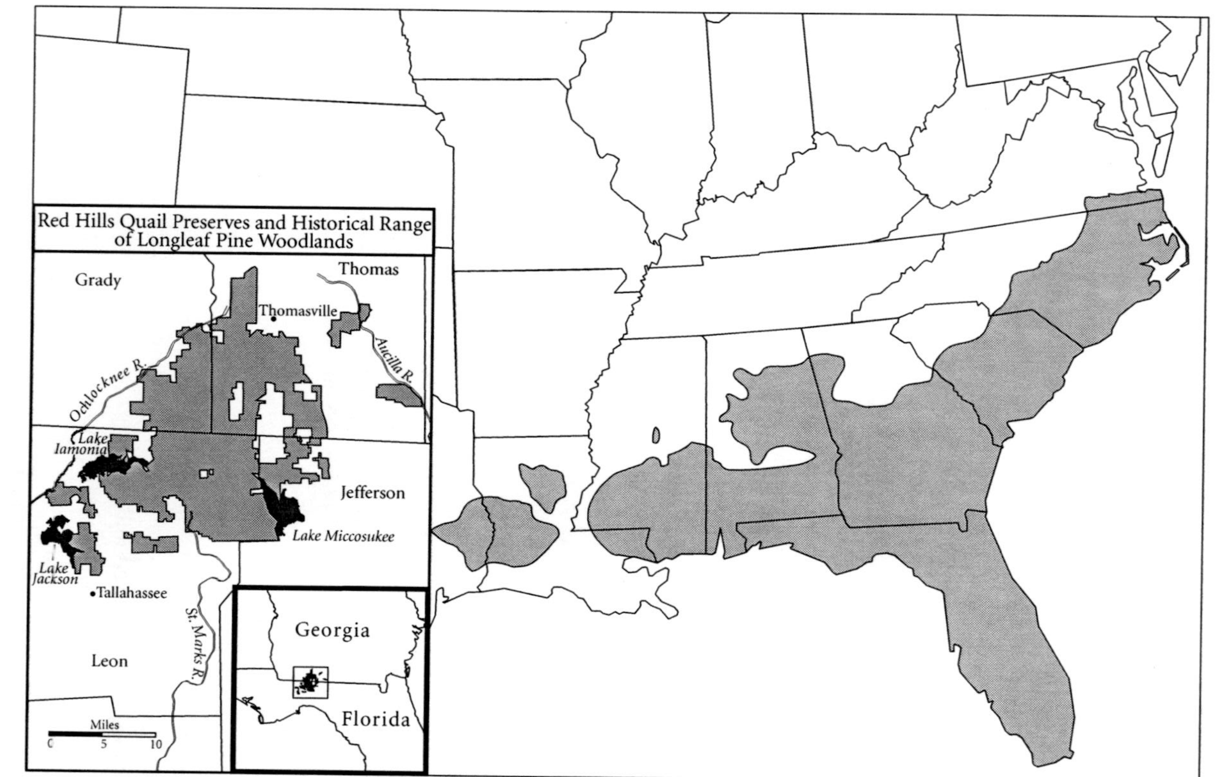

Figure 1. Red Hills hunting preserves and the historical range of longleaf pine woodlands.

that separates the coastal plain from the undulating topography of the piedmont; and it retreated at times hundreds of miles beyond our present shoreline. As the last glacial period came to an end about ten thousand years ago, the seas rose to establish the shores that we now know as the norm. These long ebbs and flows resulted in dumps of both marine deposits and piedmont runoff to create a host of different soil compositions. Soil taxonomists classify coastal plain soils, like all soils, based on the mixture of mineral and organic materials, age, relief, moisture retention, and endless other factors. Soil moisture and soil texture are the most important gradients for us laypersons, and the coastal plain formed consistent soil patterns that ranged from the driest and sandiest, known as Entisols, to the wettest and muckiest, known as Histisols. Most of the uplands, where the longleaf pine eventually reigned, fell somewhere in between these extremes. The sands and silts, some dry and some wet, predominated, with various mixtures of clayey material interspersed throughout the upland sites as well. Few of the upland soils would be memorialized for their fertility, but the vegetation that took root on them developed intricate adaptive strategies, and they formed the base for what became one of North America's signature ecological communities.[18]

Three constituent parts merged to shape the region's upland ecology: longleaf pines dominated the overstory, grasses the understory, and fire took care of everything in between. Attempting to explain the system's evolutionary ecology can quickly devolve into metaphors of chickens and eggs, but I think it is a safe bet to start with fire. More than any other natural force, fire shaped what we have come to know as the longleaf pine biome. Fire's arrival on the scene came initially as a result of lightning. The coastal plain has always been among the most thunder- and lightning-prone places in North America, and it is well known as a place that can conjure up an afternoon thunderstorm in an instant. Ed Komarek, one of Herbert Stoddard's protégés, who is also an important figure in this story, once recorded ninety-nine lightning fires in one day in Florida, which in the distant past would have generated enough spot fires to eventually merge as one sweeping line of flame across the countryside.[19] In terrain only occasionally interrupted by fire breaks—usually water courses and their less fire-prone hardwood bottomland forests, or a sudden change in topography—one lightning storm might result in a burn of millions of acres. That was before humans fragmented the landscape, but even after Native Americans arrived around twelve thousand years ago, they recognized a useful tool when they saw one. They used fire to drive game animals, to encourage the growth of forage crops, and to clear agricultural land, among other things. Colonial settlers and slaves also negotiated life with fire after they arrived on the coastal plain, adding livestock grazing to the list of activities enhanced by fire.[20] Both natural and anthropogenic fire, then, ran rampant across the coastal plain landscape for many millennia and structured the region's vegetative and animal communities into a complex assemblage of fire-loving species.

The namesake of the system, the longleaf pine, established its dominance over the coastal plain uplands in lockstep with fire and water. Its entire life cycle is structured around those two essential ingredients. The longleaf pine begins life on bare dirt recently exposed by fire, where the seed can reach the soil and begin to extend its young roots into the ground in search of water. It rains a lot on the coastal plain—about forty to sixty inches of rainfall per year depending on the location—but those sandy coastal plain soils act as a sieve through which moisture moves away from the ground surface, making a stable supply of groundwater difficult to come by for surface plants. Longleaf pines compensate by growing a long root to tap into the water source. It is the only tree in the southeast with such a taproot, which can plunge as deep as fifteen feet on the coarsest, driest soils.[21] The dense core of a mature taproot is supersaturated with oleoresin and is familiar across the coastal plain as "fat lighter" because of its combustibility and its capacity to hold a flame for what seems an unreasonable length of time.

The young longleaf pine channels most of its energy into growing the taproot for the first five to seven years of its life, while above ground it patiently bides its time in the grass stage, appearing as a clump of needles bound together neatly at the ground line. Those needles protect the terminal bud from fire, which in turn serves to protect the longleaf from being shaded out by tree species intolerant to fire. The longleaf pine loves the sun as much as it does fire and would not last long under the shade of a closed canopy. This grass stage gives the tree its reputation as a slow grower, and it does not put on much height until the taproot is safely situated near a consistent water source. Once that happens, the terminal bud begins to make steady strides upward into the candle stage, so named because the stem's new growth appears white and waxy. While in this stage, the longleaf is briefly susceptible to fire, but a scaly bark soon covers the stem, and new needle growth protects the bud again. Now it is ready for its decades-long journey toward maturity, when it reaches heights of up to 120 feet and an umbrella-like flat top slowly develops into a lovely visual indicator of old growth.[22]

If the longleaf pine dominated the canopy, the grasses came to rule the understory's vegetative communities. Indeed, this system shares as much or more with the grasslands of the prairie and plains states as it does the forests of the East. It is not even accurate to call this system a forest—it rarely achieves the necessary tree density for such a moniker. Longleaf pines most often grow in what modern systems of vegetative classification call "woodlands" or "savannahs." That means it is marked by an aesthetic of parklike openness, where sunlight can reach the ground and do its photosynthetic work. Wiregrass is longleaf pine's most frequent companion in the eastern part of its range, and the bluestem grasses take over on the western end. These are bunch grasses, so called because the leaves of individual plants grow in a bunch, not unlike a longleaf pine in the grass stage. Any number of other grasses, legumes, forbs, and shrubs intermix with the bunch grasses to form an understory with the highest plant species diversity in North America, and one that approaches the diversity of tropical forests on small scales.[23] The understory also forms a composite picture of subtle beauty and relevant function. Along with fallen pine needles, these understory plants carry out the essential job of fueling hungry fires—essential because it is a job of systemic self-preservation. The longleaf pine and its understory associates not only provide fuel for fire, they also depend on it for their survival.

Most of the animals that dwell in this system reflect its status as a fire-maintained grassland. They all subsist on the fire-adapted vegetation, whether directly as herbivores or through fuel conversion as insectivores or carnivores. And they have highly adapted strategies for evading the direct effects of fire. Of the vertebrate species found in the longleaf pine woodlands, diversity is very high for the amphibians and reptiles and less remarkable for the birds and mammals. There is, however, a high degree of specialization for all groups. Those creatures most closely associated with the longleaf pine—the red-cockaded woodpecker, gopher tortoise, flatwoods salamander, and indigo snake, to name a few—are highly dependent on the interlocking components of longleaf pine, grasses, and fire and do not survive at high rates without that combination.[24]

The longleaf pine community, though it dominated the coastal plain, was not an ecological monolith. It was part of a diverse mosaic of hardwood bottomland forests, upland woodlands, and transitional areas, or ecotones, all overlapping and intermingling, gradually giving way from one to the other. Of the region's ninety million acres, the longleaf pine uplands constituted about fifty-seven million acres, and the rest was mixed hardwood and pine of one make up or another. Even where the longleaf was dominant, its composition varied across space and time as a result of terrain, soil quality, fire patterns, disturbance histories, and eventually human land-use practices. Ecologists count no fewer than six primary physiographical regions, and at least 135 vegetation associations where longleaf pine dominates.[25] Despite so many early travel accounts that bemoan the monotony of the piney woods, the longleaf pine region was quite varied in its own subtle ways.

The longleaf system is remarkable enough for its elegant intricacies, and following the ways that Stoddard and others discovered those intricacies has been an exciting intellectual pursuit. But in seeking out those discoveries, it has been easy at times to lose sight of what is at stake here. The longleaf pine biome is one of the most biologically diverse systems in North America, and it is also one of the continent's most threatened. Healthy longleaf pine woodlands remain on only 2.2 percent of their original range, and only .2 percent saw enough fire by the turn of this century to maintain the system's historic diversity.[26] Fire must be frequent to do its job properly, and lightning fires do not race across the landscape with the frequency they once did. In fact, from the time Native Americans arrived on the coastal plain until today, the longleaf pine woodlands have become increasingly dependent on fire as a cultural technology, not as a naturally occurring force. By the mid-twentieth century, only people could consistently duplicate the natural circumstances that created and maintained the longleaf pine uplands. The longleaf pine–grassland system, then, represents a dynamic conjuncture of nature and culture, not just a static, mature wilderness. The normative environment that conservationists such as Herbert Stoddard came to advocate was a flexible, diverse landscape with a multilayered history; the disturbance of fire, as both natural process and cultural technology, was the key ingredient holding it together. So when we speak of ecological preservation in the longleaf pine woodlands, we are also speaking of cultural preservation.[27]

The following narrative leaps around from place to place and covers much chronological ground, but one place serves to bind it all together. The Red Hills contains most of the environmental characteristics of the longleaf belt at large, yet it is also a discrete subregion within that larger system. Politically and socially, the Red Hills of 1880 overlapped the boundaries of three counties: Thomas in Georgia, and Leon and Jefferson in Florida. The Georgia State Legislature carved Grady County out of Thomas in 1905, which included the western edge of the Red Hills.[28] Thomasville, Georgia, and

Tallahassee, Florida, were the social and economic hubs of the region, and they are this study's most important urban reference points. Geographically, the Red Hills subregion is relatively easy to define. Its northern and southern borders are Thomasville and Tallahassee, respectively; on the west is the Ochlocknee River and to the east, the Aucilla River. Below Tallahassee, the Cody Escarpment drops off into sandy-soiled flatwoods and savannahs that slowly descend toward the Gulf of Mexico.[29] The Red Hills' key topographical features include gently rolling upland hills, transitional bogs, streams, sinks, and bottomlands around the larger rivers. As the region's name suggests, the soils of the uplands are a mix of silts and red clays, which historically hosted a complex mix of plant communities. One study counts as many as twenty-four natural communities within a portion of the Red Hills, with the fire-maintained upland pine forest as the most common.[30]

The longleaf woodlands of the Red Hills stood as unlikely survivors after the Civil War. In contrast to most other areas in the longleaf uplands, the Red Hills grew to be a prominent plantation district before the war, a distinction that, ironically, would play an important role in its becoming a conservation landscape. Unlike so many other areas across the nation that eventually came under the control of various conservation regimes, the Red Hills was not known for its economic or agricultural marginality. The soils were rich, and where suitable for cultivation, slaves and planters cleared the land in the 1820s, 1830s, and 1840s to make way for cotton, particularly in Leon County.[31] The rolling hills, however, made large-scale cultivation untenable in much of the region. The result was a patchwork landscape of large and small fields scattered throughout substantial blocks of fire-maintained woodlands. After the war, as in other plantation districts, landowners and former slaves adopted the crop-lien system, and former slave families dispersed over the land as tenants and sharecroppers. Planters struggled to come to economic terms with losing their major source of wealth—slave labor—and croppers entered a crushing debt cycle that required an increasing reliance on cash crops.[32] Despite this structural reordering, the Red Hills continued to be an agricultural landscape, and the longleaf woodlands remained as well. In fact, many of the land management traditions of southern agriculture were responsible for the maintenance of these forests. Tenants, sharecroppers, and landowners continued the long tradition of burning in late winter and spring to rid the fields and forests of a year's worth of accumulated growth. The yearly burning habit had several practical purposes. It helped to prepare for spring planting, it regenerated succulent forbs and grasses for grazing livestock, and it helped to control pests such as ticks, chiggers, and mosquitoes. Moreover, setting fire to the brushy undergrowth made living in and near the woods easier. This cultural practice of burning, in turn, allowed the region's longleaf pine woodlands to flourish, though that was not necessarily the intent behind it. This was the unlikely confluence of ecological function and cultural preference.[33]

It was in this context that wealthy northern travelers began to value the Red Hills region for its aesthetic, recuperative, and recreational properties. Chapters 1 and 2 root this story in two corresponding trends during the decades after the Civil War: the rise of Gilded Age anxiety and Progressive-era science. Chapter 1 examines the northern tourist trade in the Red Hills and the subsequent conversion of southern farms and woodlands into private northern-owned retreats. Scholars have long noted the many important links between northern capital and the New South, but few mention health, disease, climate, wildlife, and aesthetics.[34] Whereas many northern industrialists fixed on the region for its timber, ore, soils, and plentiful cheap labor, in the Red Hills and several other areas in the South they were more interested in the healing properties of the natural environment and outdoor recreation. In searching for healthy, restorative landscapes, the new class of landowners in the Red Hills brought with them distinct ideas about nature and land use; and they effectively superimposed their ideas about the aesthetic environment over the human and ecological systems that were already in place. These notions of aesthetic nature came from urban concerns about disease, bodily health, and new threats to the entrenched power structure in the Gilded Age and Progressive era. By keeping large chunks of this landscape outside of government oversight and industrial production, they promoted ecological stability, for sure, but these exclusive estates also represented and fortified a social hierarchy.

Landowners embraced one activity in particular as representative of their social rank—hunting bobwhite quail. And when they perceived a decline in quail populations in the 1920s, landowners turned to a federal agency, the U.S. Bureau of Biological Survey, to help them manage their lands. The Biological Survey hired Herbert Stoddard to run the project, and from his arrival in 1924 until his death in 1970, Stoddard was a major voice of conservation in the South. To fully understand his distinctive intervention in the Red Hills we need to understand his background, so chapter 2 leaves the region to examine Stoddard's ascent into the role of conservation practitioner and scientific expert. It follows his growth as a child in the longleaf woodlands of central Florida, his apprenticeship as a taxidermist in the upper Midwest, his work as a professional taxidermist and specimen collector in natural history museums, and finally, his rise as an ornithologist of national prominence—all with no formal training beyond primary education. Stoddard's biography reveals the still fluid nature of science and conservation in the years leading up to and after World War I. Despite the increasing specialization of the biological sciences and the growing professionalization of government conservation departments, figures such as Stoddard remained in positions of prominence. When Stoddard came to the Red Hills as a field agent of the Biological Survey, there was still no formal route to such work; he became a professional through a series of apprenticeships, and a scientific expert through rigorous study, trial and error in the field, and fortuitous personal connections. Nor was there a codified method for carrying out research on wildlife and its habitat; as a federal employee he had a great many resources on hand to be sure, but he relied most on his own background to devise a research agenda that drew on both scientific and local knowledge. The result, I argue, was a practical system of scientific land management that reflected the contingencies of place.

Stoddard's research on bobwhite quail would have a major impact on the development of wildlife management as a distinct scientific field, and a revolutionary influence over the management of longleaf pine woodlands. Chapters 3, 4, and 5 detail the implementation and management of Stoddard's Cooperative Quail Investigation and assess its major results. Chapter 3 examines what was perhaps Stoddard's most important contribution to science and conservation on the regional level: he concluded, against prevailing and dogmatic wisdom, that the generations-old southern practice of burning the woods was a crucial component for the stability of the longleaf pine woodlands. From his rather innocent proclamations of fire's beneficence to his stern defense of his findings, this chapter charts a controversy in the piney woods that still resonates throughout the nation today. After the South's forest resources fell during the industrial cut from the 1880s to the 1920s, professional foresters entered the region for the first time with little knowledge of the coastal plain's historical ecology. They immediately attempted to snuff out the long tradition of burning the woods and in the process transformed environments on a large scale. Without fire, vast areas of the coastal plain's environment changed almost irrevocably. Stoddard's land base in the Red Hills, on the other hand, was largely protected from the industrial cut, and it became his laboratory to observe and then demonstrate the beneficial uses of fire. He was not alone in his efforts, but his landscape became one of the premier strongholds of longleaf pine remaining in the South. By arguing vehemently for the continued use of fire in the southern coastal plain, Stoddard became a defender not only of fire itself but also of a fire culture.

The second major outcome of Stoddard's study was a reexamination of the ecological relationships between predators and prey. Chapter 4 thus details how Stoddard came to many of his conclusions regarding predator and prey, and assesses his work in the context of the rising field of wildlife management. In 1930, soon after Stoddard completed his quail study, he left the confines of the Biological Survey to become a private consultant, but his national stature as a wildlife conservationist grew tremendously. In a time when the Biological Survey participated in numerous predator eradication efforts throughout the country, his work on quail and their predators led to a thorough reconsideration of the subject. Stoddard's work helped to create a scientific basis for early public opposition to the eradication programs, and it also led him to become one of a handful of figures to shape a national policy regarding wildlife conservation. On the heels of Stoddard's *The Bobwhite Quail* in 1931 came Aldo Leopold's *Game Management* two years later, and wildlife management came into its own as a viable profession for a new generation of conservation experts. By 1936 several prominent figures had formed the Wildlife Society to fill an organizational void and provide a national outlet for research and professional development. Many scientific and ethical arguments over predators and prey came to a head over the formation of the society, arguments that shot to the core of questions about culture and nature. Through these developments, we begin to see the formation of a conservation ideology that borrowed from the theoretical principles of ecological science and the everyday practice of wildlife management; the result was a pioneering template for today's conservation biology.

Chapter 5 examines this new ecological perspective within the context of agricultural landscapes. The growth of a large agricultural bureaucracy during the New Deal opened up new opportunities for the new field of wildlife management, but it also led people like Herbert Stoddard and Aldo Leopold to question the motivations and results of such expansion. The mechanization and industrialization of agriculture became a major source of contention for conservationists interested in biological resources. On the advice of government experts and industry representatives, farmers across the nation commenced an "agricultural face lifting," as Stoddard called it, and transformed the very environments that were beneficial to so many wildlife species. From their interest in animals, wildlife biologists mounted a serious challenge to agricultural intensification. On his Red Hills hunting preserves, Stoddard turned to the landscape mosaic of tenantry as a normative environment against which to measure change. Though not interested in maintaining the social arrangements of tenantry, he argued that its landscape diversity formed a type of biological reserve that would meet the productive needs of both people and nature.

The years during and after World War II were a time of major transition in both the Red Hills and the southern coastal plain at large. Chapter 6 explores the effects of the war and its aftermath on conservation in the region, focusing primarily on the increased attention to forestry on the quail preserves and the effort to balance expanding productivity with wildlife habitat. As other areas of the postwar South made a wholesale move toward industrialized forestry and agriculture, the quail preserve owners also felt increased pressure to make their land economically productive. Again, they turned to Herbert Stoddard and his associates, who by this time included Ed Komarek, Roy Komarek, and Leon Neel. Their charge was to adapt their land management into a system that met financial needs, yet maintained the desired conditions of aesthetic and ecological diversity. Stoddard and Neel's work in the forests is of particular interest in this chapter. They developed a selection-cut system of forestry and continued to apply fire in frequent intervals, thereby opening up the forest for wildlife habitat and encouraging vigorous longleaf pine regeneration. Selection forestry was nothing new, but while Stoddard and Neel devised their own variation on that theme, the pulp and paper industry was busy changing the face of southern forestry. Forestry schools, state departments, and industry foresters spread the gospel of short-rotation, industrial forestry across the South, and many practitioners characterized Stoddard's system as an anachronism in a modern world. In some ways, they were right. Stoddard's system was productive, but it was not industrial. The timber that he cut and sold moved through a system of brokers, machines, markets, and corporations, but his method of forestry itself was not a product of that systemic industrial world. It was, instead, formed at the margins of industrial forestry, a product of trial and error experimentation based on ground-level concerns for ecological health. The quail preserves, then, in some ways continued to act as refuges from modern industrialism even as it pervaded the postwar rural South. The preserves became environmental spaces on which to practice a type of forestry and agriculture that was connected to the modern industrial ideal, yet not shaped by it.

Chapter 7 wraps up the story with an examination of the first twenty years or so of Tall Timbers Research Station. Founded in 1958, Tall Timbers resulted from the efforts of Stoddard, the Komareks, Neel, and preserve owner Henry Beadel to institutionalize a Red Hills brand of conservation science. It would soon develop into a crucial voice of conservation and science in the Southeast. With the inception of their Fire Ecology Conferences in 1962, Tall Timbers became the leading voice, and for many years the only real outlet, for the dissemination of research on the ecological effects of fire. In the beginning, their focus was on the Southeast, but as Komarek grew into his role as director, Tall Timbers expanded its field of view to address fire regimes in the western United States and the world. With his focus on fire and its use to regulate natural systems, Komarek became an important advocate for the active management of ecological resources.

Forgetting all the talk of biocentrism and diversity for a moment, one wonders if practitioners such as Stoddard, the Komareks, and Neel simply liked to get their hands dirty. Their goal, after all, was production from the land, even if they measured production in ecological quality and beauty rather than economic output. The productive work of traditional wilderness preservation involved less hands-on labor than what they were doing, and even if the longleaf pine woodlands responded positively to wilderness protection, this group of land managers may not have adopted it. Theirs was a field science embedded within an agrarian society, practiced among farmers, and absorbed by the ethos of agriculture. And yet, it did not follow the relentlessly rational inclinations of modern agriculture, forestry, or other forms of conservation science for that matter. In many ways their work became marginal to the larger scientific and conservation community and was often dismissed out of hand, but it was revolutionary all the same. Stoddard's work represents a conservation alternative, and while such historical alternatives usually died a quick death, Stoddard found a niche that has largely been hidden behind the impenetrable boundaries of exclusive private landownership. Within those boundaries, Stoddard and company produced healthy ecologies using the tools of other types of productive work. *Conserving Southern Longleaf*, then, tells the unlikely story of the rise of a unique environmental ideal based in a distinct corner of the South, but with implications for the entire nation.

CHAPTER 1

From Public Playground to Private Preserve

John W. Masury, a wealthy paint manufacturer from New York, recounted his 1889 southern journey to Thomasville, Georgia, as nothing less than an ascent into the heavens. On the train ride from New York, "rain was the order all the way . . . until Thomasville was almost in sight. An hour before we reached our destination the clouds broke away and revealed the sun's face, and for sixty consecutive days 'old Sol' rose in splendor and set in glory." His stay that winter was almost Edenic. In contrast to the grubby urban environs of New York, in the countryside between Thomasville and Tallahassee, Florida, an area known as the Red Hills, one "might ride or drive or walk in the pine forests, with entire comfort and without danger to health." Masury considered these healthy jaunts possible because of the climate, which he described in spiritual terms: "There is ever about it a softness and sweetness which cannot be enjoyed elsewhere. To inhale the air there is equal to a drink of the 'nectar of the gods.'"[1]

Just six years before Masury penned his flowery prose, another visitor, this one from Louisville, Kentucky, had a very different view of the Red Hills. He reported to his newspaper audience in 1883 that "the broad, roomy mansions under the liveoaks and magnolias have disappeared or fallen into decay." Turning his attention from the landscape to the people, he described planter families who were "for the most part broken in spirit and ruined in purse by the war," and African American tenants who struggled to engage a "simple and aboriginal style of farming without the most distant idea that there are other crops than cotton or that the progress of agricultural science has developed new methods and new implements. . . . He lightly scores the surface of the ground with a plow that is but a slight advance on the pointed stick figured upon the obelisks of the ancient Egyptians."[2] What accounts for these ostensibly contradictory accounts? How could a landscape full of such obvious economic despair be the subject of Masury's effusive praise?

That Masury had just opened an eighty-room hotel in Thomasville partly explains his rhetoric, but he was not alone in singing the praises of the Red Hills. Simply put, the Red Hills environment had become healthy, at least in the eyes of wealthy travelers. During the last three decades of the 1800s, any place deemed healthy—first by the medical community, then by local commercial elites—was bound to attract the attention of an expanding class of health seekers. This new class of traveler, borne from America's booming industrial economy, headed into particular natural spaces in search of cures for all sorts of physical and psychological maladies, including tuberculosis, hay fever, asthma, and what was then called neurasthenia. Along with mountainous, coastal, and hot spring regions, the piney woods of the South were considered by physicians and health seekers to be particularly salubrious. Turn-of-the-century Americans defined health broadly, and wealthy northerners soon began traveling south for a variety of reasons loosely affiliated with restoring healthy bodies. Many had an enthusiasm for outdoor recreation in the region's fields, forests, and waterways, others a romantic penchant for the Old South and its social hierarchies. Some sought respite from cold northern winters and the emotional and physical rigors of urban life. The connecting thread to all of these motivations was the malleable notion of health. Wealthy travelers, whether suffering from bodily disease or not, wanted to find a healing landscape, one they could rely on to impart salubrity.

Changing notions of health toward the end of the nineteenth century—and the prescriptions for travel that accompanied them—had a significant influence on the environments of healthy destinations like the Red Hills. Indeed, the environment was the object of a great deal of interest because it was an active medicinal ingredient. When post–Civil War travelers began their search for healthy places, they took with them well-established medical concepts that directly linked their bodies to the environment. As historian Conevery Bolton Valencius has shown, the world of nineteenth-century Americans was not one "in which the environment stopped at the seeming boundary of the skin," and many experts and laypersons alike assessed particular environments in terms of bodily health.[3] Those assessments often resulted in the alteration of "dangerous" environments such as swamps and lowlands, but they also led increasingly to various forms of environmental preservation. By the late nineteenth century, physicians had developed a highly codified taxonomy of climate and physical terrain as it related to disease and health and advised patients to seek out those places that seemed to work best for each disease. Despite the late nineteenth-century move toward theories of disease that isolated the body from its environmental surroundings, the travel cure, as it was known, continued as a common treatment for a variety of diseases during the Red Hills' heyday as a health resort.[4]

The extent to which the perceived health of place acted as an important marker of conservation concerns plays a critical role in this story. Americans in the late nineteenth century understood and judged landscapes in terms of which ones were and were not healthy, and those judgments often translated into acts of conservation. Therapeutic remedies like the travel cure may carry little medical weight today, but as historian Gregg Mitman has argued, the search for healthy places in the late nineteenth and early twentieth centuries left a sizable ideological and material footprint in its wake and had a profound influence on local land use and environmental change. In the Red Hills, perceptions about health and environment led not only to a substantial shift in landownership but also introduced a new set of expectations for the environment. As Mitman argues, health itself became a product of nature, a natural resource.[5]

The height of tourism in the Red Hills spanned roughly from 1880 to 1920 and provides a particularly striking example of how a place of bodily health eventually became one known for ecological health. The region's healthfulness became most closely aligned with the aesthetic of its agrarian forests. The longleaf pine woodlands were of primary interest to visitors, and while not possessing the grandeur of western mountains and canyons as defined by nineteenth-century ideas of wilderness, they offered a compelling aesthetic to city-worn travelers.[6] Both modern and historical observers most often refer to old longleaf pine woodlands as "parklike," and in the post–Civil War era the Red Hills still held thousands of forested acres. Widely spaced pines towering up to 120 feet tall, a lush ground cover rarely more than waist high, and plenty of sunlight gave visitors a view through the forests unobstructed by any midstory growth. Scattered throughout these woodlands were the components of plantation agriculture, which offered another visual indicator of health. If a little derelict due to war and Reconstruction, the plantation houses, surrounding fields, and recently scattered tenant shacks were still testaments to the social and economic hierarchies many travelers thought to be under threat in an industrializing age.[7]

While local black and white southerners struggled to make a living off the land and many other residents and nonresidents eyed the vast stores of southern timber, health travelers sought relaxation, physical and psychological restoration, and outdoor recreation—all typical prescriptions for escaping the rigors of urban life. The latter was of particular interest, and travelers quickly fixated on a little bird, the bobwhite quail, as the object of recreational pursuit. For a certain set of travelers, quail hunting eclipsed all other activities, and the lands that harbored bobwhite quail became highly desirable for those who could afford them. By the early 1920s nonresidents had purchased about three hundred thousand acres of plantation land in the Red Hills—and well over two million acres across the Deep South—and converted the countryside into a group of exclusive winter hunting preserves. These spaces slowly transitioned into landscapes of ecological conservation as scientists came to recognize their biological diversity, but southern hunting preserves began as exclusive landscapes of social and cultural preservation.[8]

The transformation of the region by health tourism first required a major shift in medical thought, particularly regarding therapeutics. Remedies for diseases such as tuberculosis and neurasthenia were wide ranging, but for those patients who could afford it, physicians deemed a change in climate the most effective treatment. What was known as the "climate cure," or climatotherapy, dominated the literature on a host of diseases at the end of the nineteenth century. Flowing out of Hippocratic and Galenic humoral medicine, climatotherapy posited that certain types of air, determined by a mix of altitude, temperature, and humidity, invigorated the depleted consumptive. The climate cure, and its relative, the wilderness cure, represented a shift in thinking on the internal workings of disease that began around the mid-nineteenth century. Whereas traditional humoral medicine considered disease to be inflammatory, requiring treatments like bloodletting, physicians now thought diseases depleted vitality.[9] This shift was responsible for a therapeutic revolution. As urban Americans drifted away from their agrarian roots, sufferers traveled into the countryside to infuse their bodies with vitality through physical activity and a change in climate. According to historian Georgina Feldberg, late nineteenth-century "physicians associated consumptive disease with a changing social order, [and] their therapeutic advice to consumptives attempted to preserve and re-create the world that they feared was slipping away."[10] Early on, such therapy was little more than an informal jaunt through uninhabited space, but by the 1870s and 1880s the medical establishment developed a more codified taxonomy of climate as it related to disease, thus leading to the development of health resorts in the allegedly most salubrious places. Due to its mild climate and its relatively nonindustrial landscape, the South was particularly attractive.[11]

The post–Civil War South was brimming with wealthy northerners looking for healthy landscapes. Though the Red Hills region had much to recommend it, the health trade seemed just as likely to bypass the area as to transform it. First of all, promoters of Thomasville and Tallahassee had a great deal of competition. Places such as Pinehurst, North Carolina, and Aiken, South Carolina, like Thomasville, offered their pine forests as purifying filters for miasmic air; Asheville, North Carolina, situated high in the Appalachian Mountains, advertised its dry air as a salve to aching consumptive lungs; and the therapeutic hot springs of Arkansas and the Virginias were a major draw for rheumatoid arthritics. All offered physical and spiritual rejuvenation to physician and patient, as well as the recreational traveler. In addition, the sun-speckled coasts and wild interior of peninsular Florida also beckoned. Despite being the site of the continent's oldest European settlement, vast stretches of Florida still seemed void of human improvement, just the sort of unconfined wilderness that increasingly appealed to health tourists. The Red Hills emerged as a health retreat, then, in competition with many other places making similar claims.[12]

Almost immediately after the Civil War, travel writers flooded northern markets with tracts about the South. Florida was a particularly popular destination, and local leaders in Thomasville and Tallahassee soon realized an opportunity to capture some of that southbound traffic. Writers like Ledyard Bill, Daniel Brinton, George Barbour, and Sidney Lanier wrote glowingly of the peninsula's warm winter air, especially in regard to pulmonary tuberculosis. By the time Brinton published his *Guide-Book of Florida and the South* in 1869, "even those who lay no claim to medical knowledge are well aware how often the consumptive prolongs and saves his life by a timely change of air."[13] This is not to say that all southern air would put you back on your feet. Disease was still believed to be environmental in nature, and until the acceptance of the germ theory of disease, it was still supposed that the body reacted to specific environments in predictable ways. The Floridian lowlands, for instance, were long thought to be a pestilent, malaria-ridden expanse of muck. Brinton recognized the region's threat to the body, and counseled that seasonal timing was key to southbound travelers. Come too early and they might encounter "the swamp miasm [that] begins to pervade the low grounds, and spreads around them an invisible poisonous exhalation."[14] Miasmic air was a very old threat, one that had sent antebellum coastal planters fleeing for the piedmont and mountains during the summer. The lowland South was infamous for exuding noxious air thought to cause a host of diseases, including malaria and yellow fever. So as Brinton's comments indicate, a shift in medical thought did not suddenly make the South healthy. Miasmic air continued to flow in the summertime, but with cold weather the miasm "loses its power," and only after "one or two sharp frosts have been felt in New York or Philadelphia, [is] the danger chiefly past."[15] The winter traveler, then, had few climate-related worries.

If the traveling invalid did choose the South, that was only the first of many decisions. A simple change of air, any air, was an improvement over stale city air, but as climatotherapists developed their trade, the decision of which climate to choose became "a question of vital importance. An error here is fatal," according to Brinton. If, for instance, the consumptive followed friends or fashion to the crisp mountainous air of Asheville, North Carolina, when the disease called for the balmier atmosphere of St. Augustine, "he goes at his peril. . . . There are some whose safety lies in the mountains, others who can find it nowhere but on the sea shore."[16] Dr. Charles J. Kenworthy, president of the Florida Medical Association, advised that "facts, figures, experience, and favorable factors of climate" should be tailored to each individual patient.[17] In fact, there were about as many therapeutic locations as there were maladies, each with its own elaborate justification as the most suitable place to begin the healing. Many physicians and travel writers engaged in a type of regional promotion that skewed evidence in favor of their own region, but others could not point a patient in any one direction with much confidence no matter how elaborate the reasoning. Even Brinton, after matching a long list of constitutional complaints with particular climates, threw up his hands, saying: "Have you a fancy for any particular spot among those famous for salubrity? Is there a pastime or pursuit to which you are addicted? Do you love to boat, fish, hunt, ride, camp out, botanize, photograph? Indulge your taste. Such considerations have quite as much weight as many a medical reason."[18] Indeed, recreational pursuits would soon take precedence for the majority of winter travelers.

As early as 1869, a few years before Henry Flagler and Henry Plant created their tourist wonderlands on the Atlantic and Gulf shores, northerners were already indulging their tastes in the warmth of Florida. Ledyard Bill reported in that year that it had already "attracted considerable attention as a winter resort for invalids and pleasure-seekers. . . . Visitors to the State are already numbered by thousands, and each year since the war has witnessed a rapid increase."[19] The most common entry point was Jacksonville and the St. John's River. From there, most visitors traveled upriver to coastal St. Augustine and the Indian River section, or into the less inhabited areas farther south. Some, though, preferred the older, more cultivated countryside around Thomasville and Tallahassee. As an interesting side trip, or as a layover on the way to Florida, the Red Hills drew growing numbers of curious health travelers looking to discover what they considered to be the Old South. And when local commercial elites in the Red Hills noticed more northerners passing through, they also began to conjure up ideas about holding them there. Both Thomasville and Tallahassee were relatively well connected to the East by rail, though not to each other. By the late 1870s, Thomasville had particularly good rail connections, with lines running to Savannah and Albany, both of which connected to points northeast, as well as the booming Ohio and Mississippi Valleys.[20]

Although the rail system brought them to the Red Hills, it was the region's uniqueness, both environmental and social, that captured the northerners' prolonged interest. Early northern visitors clearly considered the Red Hills a distinct landscape compared to the flatwoods and coastal environments farther south. George Barbour noted in 1882 that his trip up the St. Johns River allowed him to see "the wilder and more remote regions," while his visit to the Red Hills gave him "an opportunity to learn of the older and more populous sections." "On every side in all that region," he wrote, "were seen large old plantations . . . giving evidence of a long-settled region."[21] Nature writer Bradford Torrey was thrilled to leave "the monotony" of the St. John's hinterland. As he approached the eastern edge of the Red Hills,

> there came a sudden change in the aspect of the country, coincident with a change in the nature of the soil, from white sand to red clay; a change indescribably exhilarating to a New Englander which had been living, if only for two months, in a country without hills. How good it was to see the land rising, though ever so gently, as it stretched away toward the horizon! My spirits rose with it. By and by we passed extensive hillside plantations, on which little groups of negroes, men and women, were at work. I seemed to see the old South of which I had read and dreamed, a South not in the least like anything to be found in the wilds of southern and eastern Florida; a land of cotton, and, better still, a land of Southern people, instead of Northern tourists and settlers.[22]

In the Red Hills, Torrey and other northern health seekers not only discovered a unique physical environment; they also found what they considered an authentic cultural landscape. This was the landscape many visitors came to see—not an unpeopled wilderness, but a picturesque landscape of the mythologized southern past embedded within the natural world. As northerners witnessed the industrial transformation of their own region, in the South they found the opposite, a land, according to George Barbour, "arrested in its growth, and in a state of suspended animation."[23] In reality, of course, the agricultural South was anything but suspended, but such a sense of arrested growth was essential to travelers. Making the southern fields and forests a place of health forced both travel writers and visitors to naturalize and objectify these surroundings as something separate—they created a landscape aesthetic that had little regard for the currents of history, economy, or culture. By turning sharecropping and tenantry into a static, ahistorical system, they followed the lead of many other observers, both northern and southern.

Even before the Civil War, the elite white classes of the Red Hills were in the process of defining their identity as part of a timeless plantation South. Settlers came to Thomasville and Tallahassee in the 1820s and 1830s from other seaboard states to establish a new frontier in the Red Hills' piney woods. Both Thomas and Leon counties attracted large planter families, but as historian Edward Baptist has shown, building the plantation economy was a piecemeal process constantly interrupted by disease, war, financial bankruptcy, and an uncooperative environment. By the eve of the Civil War, though, life and labor seemed so deeply entrenched for both residents and visitors as to give the appearance of an older settled land. Thomasville, Tallahassee, and Monticello were thoroughly established as county centers, and total population in 1860 grew to 10,766 in Thomas, 12,343 in Leon, and 9,876 in Jefferson County. African American slaves outnumbered whites 21,707 to 11,278 in all three counties. The majority of farms ranged anywhere from fifty to five hundred acres, but there were at least fifty-seven plantations of over a thousand acres, most of them located on the good soils of the Red Hills portions of each county.[24] These plantations—in Leon County in particular—dominated much of the landscape. As in so many other plantation districts, by 1860 local elites adopted a view that positioned slavery as a natural system holding all of society's parts in place. According to Baptist, the "planter class redrew their particular corner of a very new South as one that was 'Old,' unchanged from the past and unchanging in present and future"[25]

After a period of turmoil during the Civil War and Reconstruction, the process of remaking the new into the old began again in the Red Hills. In the historical context of slavery it was not much of an ideological stretch to naturalize tenantry into the landscape. Neither Bradford Torrey nor George Barbour distinguished much between hill, dale, field, forest, mule, or plow; they were all part of a picturesque landscape whole, a cultural landscape that travelers relished. Along with the aesthetic beauty of the natural scene were white planters anxious over losing much of their wealth with the fall of slavery, and poorly capitalized black tenants with little more than their freedom. The organization of labor that emerged, as will become clear, was a new system of debt peonage based on old hierarchies of racial, political, and economic control; and it created subtle but important changes in the landscape as well.[26]

While other post-Reconstruction southern locales set about enticing northern capital with the promise of cheap labor or of inexhaustible extractive resources like timber, local elites in the Red Hills and other healthy places recognized economic opportunity in the seasonal migration of tourists to the South. For Red Hills locals to engage the health travel market, though, they needed to make the area more than a layover point or side-trip destination. Thomasville was far ahead of Tallahassee in capturing tourists. Following the boosterist ideas of Henry Grady and his ilk, a small group of Red Hills elites encouraged local citizens to shape up and take advantage of the economic opportunity sure to be found in tourism. John Triplett, the editor of the *Thomasville Times*, was one of that town's more visible health-trade recruiters. In 1873 he observed that "the tendency is growing stronger every year, among northern tourists, to stop short of the humid atmosphere of Florida, and take advantage of the high pine lands of South-west Georgia, which affords relief in pulmonary diseases that no other section does; whilst as a summer resort, this point might, if the proper accommodations were offered, secure a large number who annually leave the low country."[27] Thomasville had several boarding houses, but nothing that could cater to finer tastes. Planter Thomas C. Mitchell stepped up to the civic plate to offer such accommodations. After a quick gestation and construction period, he completed Mitchell House in March 1875, to much praise from local media. For the *Weekly Floridian*, "the best thing about the whole business is, that the lot was bought and the hotel is being built by an old planter of Thomas County with his own capital!"[28] Mitchell did indeed build it from his own means, but he called on experienced northern hotel management to lease and run the hotel. By this time the summer health resort industry of the Northeast had many years experience and a well-developed network of managers for Mitchell and his partners to tap. Massachusetts hotelier C. S.

Sanderson signed a five-year lease on Mitchell House in 1876, and A. L. Fabyan, "one of the most accomplished hotel men of this age," was the first in a long line of northern managers.[29] While this was not the exact type of enterprise envisioned by New South architects, its origins in the commercial milieu of town life, and its collaborative nature between North and South, was typical of other more industrial pursuits in the region.[30]

Triplett was prepared to solicit Mitchell's putative customers, but the question of therapeutic legitimacy remained. It was not enough for a newspaperman to speak to the healthfulness of place; that was the realm of the physician. While construction neared completion, Dr. Thomas S. Hopkins produced an elegant argument supporting the healthfulness of the Red Hills. Hopkins, a Thomasville native and recent mayor, was active in the Medical Association of Georgia and a founding member of the South Georgia Medical Society. In an 1874 address to the Medical Association of Georgia, he expressed doubt that "there is on the globe any region of country, of the same extent, more exempt from all diseases of the respiratory organs," as was the region between the Altamaha and Flint rivers of south Georgia. In his thirty years experience he rarely found a case of pulmonary tuberculosis in the region that could not be "traced to hereditary transmission," and it could never "be attributed to climatic influence." He recommended the entire region as therapeutic, but if forced to choose, Hopkins "would select Thomas County, and preferably the town of Thomasville, on account of its elevation, its thorough natural drainage, its pure and delightful freestone waters, its dryness, its equability of temperature and its remoteness from the sea." Perhaps most importantly, Thomasville was situated "in the midst of a vast pine forest of almost unlimited extent." With such a buffer of pine acting as a filter, "the winds from the ocean reach it sifted of all saline vapor and moisture, comparatively warm and innoxious." The association quickly adopted a resolution to "earnestly and fully endorse the opinions and statements" in Hopkins's paper, and "in view of its importance to the whole country, desire to give to it the widest possible publicity."[31] Almost a decade later, none other than booster extraordinaire Henry Grady wrote, "Dr. T. S. Hopkins has been a pioneer in setting forth Thomasville's claims, and to him, as much as any man, is due her present pre-eminence as a health resort."[32]

With this focus on health tourism, Hopkins, Triplett, Mitchell, and others clearly hoped to boost their town out of its postwar economic miasma, but they did not have to corrupt the scientific literature of the time to do it. They simply capitalized on it. Hopkins apparently followed the literature closely, seizing on themes that centered on climate and trees as purveyors of health. For the Red Hills, the therapeutic keys consisted of a piedmontlike topography nestled within the warmer coastal plain, and the "vast pine forest of almost unlimited extent" that surrounded Thomasville. Unlike the Leon county portion of the Red Hills, which had much more agriculture, Thomas County was indeed covered with forest. The 1880 census shows that Thomas County had over 200,000 acres (60 percent of total) of forested land, whereas Leon County had only 56,107 forested acres (30 percent of total).[33] So when trees became the subject of much therapeutic talk, Thomasville was in a unique position to capture tourists.

As the numbers show, the Red Hills pine forest was not of "unlimited extent," but it was significant, particularly in the southern half of Thomas County. In this forest lay the Red Hills' value as therapeutic sanctuary. The value of trees as purveyors of health was a very old theme in medical literature, most likely beginning with Pliny the Elder's belief that the more resinous trees "are beneficial to consumptives," and it gained increasing medical legitimacy in America around the turn of the nineteenth century.[34] When Samuel George Morton—an American physician, naturalist, and architect of scientific racism—published his treatise on pulmonary tuberculosis in 1834, he gave full credence to the tree theory, writing, "Experience has amply proved that a dry air, in conjunction with the aroma of pine forest, is most congenial to delicate lungs."[35] Following the view that diseases like malaria and consumption were miasmatic—that is, they were atmospheric, with the air itself transmitting the disease—practitioners and theorists began to believe trees like pine and eucalyptus could both filter and counteract the offending maladies. As historical geographer Kenneth Thompson has shown, by the mid-nineteenth century the "belief in the therapeutic and prophylactic value of trees and forests . . . was firmly implanted in both lay and medical opinion."[36]

This shift in thinking about the healthy effects of trees led to increasing concern in the late nineteenth century, especially among physicians, that industrial scale timber cutting might create serious public health problems. As the northern timber industry began to make inroads on the southern longleaf woodlands, Hopkins's paper became more than a local promotional tool. In 1875 the Georgia Board of Health followed up on Hopkins's address to issue its biennial report, titled "The Influence of Trees on Health." Investigator Benjamin M. Cromwell argued that the prevalence of trees influenced the climate of any given locality, and attempted to gauge the "specific influence they exert by means of the odorous emanations they give off from their leaves, bark, wood or gum."[37] Of the various species available in Georgia, he knew of "none that are indigenous, more worthy of mention, than the common pine of our forests. . . . Besides, its merits have been ably brought before the medical profession of Georgia, by Dr. T. S. Hopkins, of Thomasville."[38]

Cromwell, though, went further than Hopkins. He not only championed the "odorous emanations" of pines but also linked them to the quality of water and soil. Like many other physicians across the United States, he recognized a vital link between environmental conditions, human land use, and human health.[39] In what can be read as an early plea for forest conservation, he explained the "silent and unobtrusive agency" of trees in "keeping up the springs, streams, and water courses of a country and thus maintaining its water supply." They also acted as sponges to drain standing water, and screens from "poisonous emanations generated to the windward of them."[40] As the industrial deforestation of the South geared up, Cromwell appealed to discretion in favor of health.[41] Trees, he argued, protected air and water, two basic elements of life that could easily go bad. Just as the piney woods of the South became recognized nationwide as a place of health, Cromwell feared that the ravages of industrial timbering would transform the region into its former miasmic self. He concluded almost desperately: "Instead, there fore, of regarding forests as encumbrances, to be got rid of as expeditiously and as cheaply as possible, to make way for the plow, would it not be better for us to exercise some discrimination in removing these valuable coadjutors of health and guardians of the fountains?"[42]

The idea that leaving trees standing would secure health was a powerful one for many. Laypersons in medicine and natural history even commented on the correlation among standing trees, well-drained soils, and health, all the while maintaining a sense of awe at the aesthetic of the surrounding forests. One visitor related to his fellow Chicagoans that Thomasville "is located in the highest and driest part of the 'uplands' of Southern Georgia, in the very center of an immense pine forest, the health-giving qualities of which are so valuable that to over-estimate the potency of their remedial virtues would be well nigh impossible."[43] Another effused, "The health-giving breezes that sweep through the pines are nowhere more delicious."[44] Robert Koch's 1882 discovery of the tubercle bacillus would eventually revolutionize medicine, but practical treatment options for diseases such as tuberculosis would remain environmentally based many years to come. The local literature continued to play up the health angle, and visitors continued to judge these landscapes based on their health-giving properties through the turn of the century. An 1898 pamphlet, "The Great Winter Resort among the Pines," reported that the "ablest physicians and specialists have for many years recommended a residence among the pines of this section, as being most beneficial to parties troubled with weak lungs or bronchial affections; and the thousands who have been cured here bear testimony to the efficacy of this climate in such cases."[45] Amid the praise, there was also concern for the region's forests. One Thomasville native, S. G. McClendon, lamented in 1889 that south Georgia's pine forests were "steadily growing smaller under the ravages of the axe," adding that deforestation "would be an incalculable disaster to the human race."[46] McClendon's apprehension was as much about economic health as bodily health, but his views echoed those of physicians. Many in the Red Hills, locals and visitors alike, agreed, and health tourism was an essential component to the survival of many longleaf pine–savannah woodlands.

Within a few short years, Thomasville was among the finest resort destinations in the South. By 1891 another visitor from Chicago could write, "Prominent Northern capitalists are investing large amounts of money there, as they consider it the most beautiful and prosperous health resort in the South."[47] Besides Mitchell House and a number of smaller hotels, a group made up of local and northern investors completed the 160-room Piney Woods Hotel in time for the 1885 season, and in 1889 John W. Masury opened his 80-room Masury Hotel. Around November every year, locals began to look for migrating visitors from New York, Boston, Philadelphia, Chicago, and Cleveland among other northern cities. Mitchell House burned in 1883, but its 3,194 registered guests the previous season were evidence enough of demand for reconstruction. It reopened in February 1886.[48] Tallahassee and Leon County felt the winter influx, too. In 1894 Bradford Torrey proclaimed that "it was exactly what I had hoped to find it: a typical Southern town; not a camp in the woods, nor an old city metamorphosed into a fashionable winter resort; a place untainted by 'Northern enterprise,' whose inhabitants were unmistakably at home."[49] Tallahassee did not have the lavish hotels of Thomasville, but in Torrey's eyes it stood as the authentic South, a characteristic that was fast becoming as commoditized as health.

The entrepreneurs of the Red Hills were not presenting a backward-looking Confederate stronghold, however. Like other southern destinations, Thomasville's tourism industry played an important role in postwar reconciliation. John Triplett, for instance, rolled out the welcome mat, imploring

Figure 2. Piney Woods Hotel, ca. 1900. The Piney Woods Hotel was one of several large hotels in Thomasville that catered to Northern tourists. The Savannah, Florida, & Western Railroad ran the special train from Savannah that dropped passengers at the front door and picked them up at the end of the season. Courtesy of the Thomas County Historical Society.

northerners to "see a live, progressive town, and meet with people who are not repining or looking back over the desolate wastes of the past, but a people who are looking hopefully forward." Visitors would still receive "a hearty, old fashioned, cordial southern welcome," but talk of past sectional conflict would not arise.[50] Triplett insisted that the citizens of Thomasville were very much in tune with the national project of reconciliation: "There are two things which are never asked about, religion and politics. Every one is left free to entertain his own views on these questions without criticism."[51] If the Red Hills was to sustain itself as a winter haven for northerners, the nation's past sectional conflicts would need to be soft-pedaled or swept under the rug altogether. But the rhetoric of reconciliation did not always prevail in more private venues. James Brandon expressed genuine surprise that his aunt took northern boarders. Because of her "implacable resentment and intense hatred of all Yankeedom," he "would advise those boarders to keep quiet on the subject of the late war and the kindred one of politics in her presence." Any breech of etiquette might cause her to "'boil over' some day and pour forth the vials of her wrath [into] the devoted heads of her amazed boarders."[52] Despite such unreconstructed feelings, the arrival of tourists in the early days often elicited little more than a shrug. Even when James Brandon's father, David, observed "a full supply of Yanks on hand" there was little to get excited about: "The town is dull—business unusually so."[53]

Most locals apparently toed Triplett's line in public at least, for northern visitors thought of the Red Hills as nothing short of an Arcadian paradise. From the northerners' perspective, the landscape itself—along with its health-giving properties—made reconciliation a much more pleasant venture. Dr. John T. Metcalf of New York wrote the *Boston Medical and Surgical Journal* to say "I wish more Northern doctors knew what I know from a series of years, of this wonderful corner of the vineyard! . . . What would you Bostonians have said had you seen us lying on my big wagon-robe, *al fresco*, at noon, whilst taking our bit of luncheon."[54] Visitor G. Q. Colton reported to the *New York Times* that Thomasville was "situated on a high belt of land 450 feet above the level of the Gulf of Mexico, and is surrounded by immense pine forests, so that the atmosphere is constantly impregnated with the aroma of the pines."[55] William Drysdale echoed Colton by elaborating on what "nature has done for Thomasville. . . . Every breeze that blows from the south or from the east must come through from fifty to two hundred miles of wholesome pine forest, and every wind from the north and west is sure to be healthful and braceful."[56]

As was the case for the majority of visitors, Metcalf, Colton, and Drysdale were not invalids; they were, according to one writer in *Popular Science Monthly*, members of "the numerous and increasing class of well-to-do, leisurely, and healthy people who seek a change of climate purely as a matter of personal enjoyment. . . . They constitute the great mass of the patrons of Southern winter resorts."[57] Indeed, health tourism catered to both the wealthy and the sick, two ostensibly separate demographics that were actually indistinct. On "the question of money," physician Daniel G. Brinton warned that the healing would only begin if sufferers were able to free their minds from the trappings of work: "If you carry the cares of business with you; if you have to pinch and spare on your journey; if you are worried about your expenses, the trip will do you little good."[58] The healing that places like Thomasville offered was clearly the domain of America's new business and industrial elite, a class that experienced its own unique health trouble around the turn of the century.

As the germ theory of disease emerged in the 1880s and it became clear that people contracted tuberculosis from one another, the closed-off sanitarium took over as the consumptive's destination. The middle- and upper-class urge to flee the city, however, held strong, and they did not have to look far to find an appropriate medical justification for doing so. The literature brimmed with descriptions of new ailments born of the industrial age, most of which were eventually consolidated as neurasthenia. Physician George M. Beard first identified neurasthenia in 1869, and he outlined its causes, symptoms, and cure in his classic medical text, *American Nervousness* (1881). He portrayed it as "a lack of nerve force," which was expended not through overstimulation of the body, but of the mind. A difficult disease to pinpoint, neurasthenia, according to Beard, included a broad constellation of symptoms—dyspepsia, insomnia, hysteria, asthma, hot and cold flashes, premature baldness, exhaustion. The common thread, though, was the correlation between modern industrial growth and increased nervousness, particularly in the classes that placed the "labor of the brain over that of the muscles."[59]

Travel writers of the time followed—and sometimes preceded—Beard's descriptions and remedies for urban diseases like neurasthenia. Daniel Brinton, writing before Beard gave a name to the disease, called it paresis, but his descriptions were indistinguishable from Beard's. Brinton took it as seriously as consumption, writing that paresis manifested itself as a state of "nervous and mental exhaustion, consequent on the harassing strain of our American life, our over-active, excitable, national temperament."[60] Ledyard Bill, too, wrote that such external excitement required serious attention, and thought the South to be an ideal place for "those who seek rest and recuperation from the steady and exacting demands of business. There is needed among those who fill the various professions more of rest and play than they get."[61] Following a narrative convention of the day, Bill passed on the experience of a contemporary who had recently been "snatched away, in the meridian of life, from over brain-work. . . . It is this over-worked class, as well as the invalid, who need to go to Florida."[62] Neurasthenia was due entirely to the external pressures of the city, and Brinton characterized it as "a new disease, a visitation of nature upon us for our artificial, unquiet lives."[63]

Escaping the artificiality of urban life was of critical importance in the development of health resorts and the later conservation of the Red Hills' woodland landscape. Brinton almost characterizes nature as exacting revenge on the wayward urbanite for creating something as unnatural as the industrial city. But in reconciling with nature, the urbanite may be able to salvage the loss of vitality. In instructing his readers to go forth into the wilds of the

South, he tells them to face their opposite. But more than wilderness, travelers were looking for a rural time and place they thought to be vanishing. This theme pervades most of the early travel literature on the Red Hills as well. The influence of diseases such as neurasthenia, and the ideas about the city and countryside that informed their construction, provide one framework through which to understand travelers' activities and perceptions in and of places like the Red Hills. Even after the climate cure fell out of favor in regard to diseases like pulmonary tuberculosis, Thomasville's promotional literature still stressed the region's restorative qualities, but it became more about escaping the demands of an urbanized modern America. In its towering longleaf forests and quaint farms, the Red Hills possessed the restorative aesthetic of nonindustrial production that fleeing urbanites demanded.

Under the influence of this new type of health seeker, by the 1890s the Red Hills became better known for its scenic drives and its sport in quail hunting than for its direct therapeutic effects on disease. Thomasville's hotels sported a variety of entertainment, including baseball, tennis, progressive euchre, billiards, dances, and concerts. Three particular activities, though, showcased the Red Hills countryside and reflected well the disposition of most visitors: lounging, driving, and hunting. The twenty-foot-wide verandas of the Piney Woods Hotel gave visitors ample space to lounge in the sun, thus escaping the demands of the business world. The Piney Woods also faced a patch of ground that impersonated the parklike aesthetic of the Red Hills' forested countryside. Known as Smith's Grove or "Yankee Paradise" until the city of Thomasville purchased the sprawling tract in 1889, Paradise Park was the center of tourist social life. During the winter months it teemed with northerners variously strolling, chatting, or stretched out for an afternoon siesta. Underneath towering old-growth longleaf pines was a well-manicured lawn with ample seating and crisscrossing pathways. The bond issue to pay for the park included provisions not to cut trees or erect buildings, a nod to the area's health-giving aesthetic.[64]

Touring the countryside in horse and buggy was a particularly popular activity. With good roads extending in all cardinal directions, as well as a fifteen-mile loop at a two-mile radius from downtown—completed in 1891—the opportunities to survey Red Hills life were ample to the curious visitor. Known as Sanford Boulevard, and later renamed Pine Tree Boulevard, the loop was especially attractive. One observer called the drives "delightful, [with] *good roads leading in every direction.*" This was something that "should be very distinctly emphasized, for herein lies the chief charm of Thomasville as a place of resort for tourist, pleasure and health seekers. . . . *No other resort in the entire South has such drives.*"[65] Another patron of the Piney Woods Hotel, A. F. Boynton of Pennsylvania, was a "most liberal patron of the livery stables," and kept a log of his driving days. Boynton spent the winters of 1888–93 in Thomasville, and out of a possible 329 days, excluding Sundays, he and his wife went driving 295 of those days. One local real estate agent, E. M. Mallette, reckoned there was no other "place in the South where one could drive an equal number of days during the winter; certainly not over the same smooth roads, penetrating the pine forest in every direction."[66] Good roads, and the scenery to go along with them, were of highest importance to visitors, and their popularity suggests that the appeal of the Red Hills was in a settled character that retained much wildness. In the horse and buggy they would not have to leave behind the domestic comforts they came to appreciate about industrialization, but they could still enjoy the scenery and fresh air of the forested landscape.

These buggy rides serve as a reminder that most upper-class travelers were already in the process of separating themselves from urban life back in their home regions. As concerns about urban health sent travelers in search of healthy places in the South, in northern cities landscape architects such as Andrew Jackson Downing, Frederick Law Olmsted, and Calvert Vaux were busy building naturalistic parks and bucolic suburbs, blending the domestic comforts of urbanity with the pastoral scenery of the countryside.[67] The curving road, as opposed to the rectangular grid, was a central component to the new landscape architecture, and the carriage ride a crucial part of the suburban and park experience. In their penchant for cruising the open forests of the Red Hills, northern travelers brought a part of their northern experience to the South. These travelers were not looking for the primeval experiences of those who went into the wilds of Florida; instead they wanted a relatively developed infrastructure, not unlike their own suburban enclaves, to ease their penetration of the region's forests. The roads they traveled in the Red Hills, however, did not pass through the manicured landscapes of bucolic suburbs or urban parks, or an untended wilderness. These forests and fields were economically productive, yet ecologically diverse, places of work.

Hunting was by far the most important activity for visitors. It gave them an opportunity to interact directly with nature and to restore the ties they thought modern society had severed. Throughout America in the late nineteenth and early twentieth centuries, hunting acquired a revered status

Figure 3. Pinetree Boulevard, ca. 1900. Pinetree Boulevard was Thomasville's perimeter road, a popular attraction for its scenic views of longleaf pine woodlands. Courtesy of the Thomas County Historical Society.

among recreational pursuits. Many groups undoubtedly considered it popular recreation throughout the 1800s, but after the Civil War, and especially after the American interior opened to development, middle-class urbanites flocked to the woods as interested in finding their ancestral selves as they were in finding game.[68] While the booming industrial economy pushed the nation toward a modernity that retained few links to its agrarian past, there was a reactionary move to restore those ties. Supporters expressed their views in a number of ways. Some hunters adhered to the diffuse back to nature movement, others to a primitivism that shed all ties to modern life.[69] Teddy Roosevelt's admonishment to cast off an effete, overrefined society in favor of the "strenuous life" also motivated many to recapture their manhood from the feminizing ways of the city.[70] Organizations like the Boone and Crockett Club, and periodicals like *Forest and Stream* and *The American Sportsman*, cropped up to define the ethical and moral standards that distanced this new class of sportsman from the less savory traditions of pot and market hunting. In doing so, elite hunters joined with a growing American conservation movement that alerted the nation's public to a dwindling supply of natural resources. But, as historian Thomas Dunlap has argued, the root of the sportsman conservation effort was as much about a threat to American hierarchies in an industrializing world as it was a threat to nature itself.[71] The growing influence of new immigrant groups and the sustained dissent of racial minorities signaled a new threat, and, as several historians have argued, sportsmen groups and landowners began lobbying state and federal governments to tighten control over the nation's fields and forests.[72]

Travelers to the Red Hills were no doubt troubled by these trends, and what they found by way of southern society may have helped to bolster their spirits. As timber interests were discovering in other areas, the South's land was cheap, its resources abundant, and its people at turns both accommodating and powerless, thus making exploitation of any kind a relatively smooth venture. Instead of resource extraction, though, visitors to the Red Hills looked to secure sanctuary from the forces of modern life. In doing so, these architects of modern American production and consumption—the forces then transforming so much of the countryside—also preserved some of the last remaining longleaf pine woodlands in the Southeast. But more than a passive act of preservation for nature's sake, the sportsmen's concern for their natural surroundings, and their apprehension over fluid social hierarchies, materialized on the ground in the form of a desired aesthetic—an aesthetic based on pre-industrial natural and cultural conditions. Small-scale peasant agriculture, mixed with the towering canopies of open forests, produced the scenery in which to pursue a civilized, artful hunt. They had little understanding of ecological processes—or that their aesthetic tastes would soon lead to a greater understanding of those processes—but they did know that their preferred game animal, the bobwhite quail, thrived in the agrarian woodlands of the Red Hills.

The bobwhite quail had a particularly lofty status in the game animal hierarchy of the late 1800s. Known variously as partridge, quail, bobwhite, or *Colinus virginianus* to the scientists, they are ground-dwelling birds that only take to the wing for short distances when flushed. During the spring and summer months they split into pairs to mate and tend nests, and they then reconvene in the fall to form coveys of five to thirty individuals. Although slaves, yeomen, and planters commonly hunted and trapped quail before the Civil War, the modern form of quail hunting—using dogs to locate and flush the coveys—followed the widespread availability of breech loading, hammerless shotguns and the proliferation of new dog breeds in the 1870s and 1880s.[73] Though the range of the bobwhite quail spreads throughout the eastern United States, only in the South did it become a source of regional identity. The planter class, in particular, embraced quail as a symbol of perseverance in changing times. Quail adapted to the environmental upheaval of Civil War, Reconstruction, and the transition to tenantry, not just surviving but thriving. Southern planters considered such survival during hard times a suitable nature metaphor for their own social, political, and economic purgatory.[74]

Although southerners gave the bobwhite quail anthropogenic traits, and they felt strongly about its identity as a distinctly southern bird, it was the northern visitors to the Red Hills who first borrowed from the ritualized hunting traditions of English nobility, thus turning the southern quail hunt into the most distinguished of American field sports. And the environs of the postbellum Red Hills were ideal for both quail and quail hunting. As a species that thrives on early successional habitat—plant communities that recolonize land after ecological disturbance—quail proliferated in the postbellum Red Hills. Three distinct habitats were of particular import for quail: field edges, old fields, and open longleaf pine woodlands. By the 1880s, as planters and laborers together forged the tenant system, dividing up expansive fields into smaller plots, field edges became far more abundant. In Leon County, for instance, the number of farms jumped from 319 in 1860 to 1,789 in 1880, and 2,428 in 1900. In Thomas County, there was a similar leap:

299 in 1860, 1,588 in 1880, and 3,183 in 1900.[75] Such disparate figures can be misleading—they reflect the new census-taking methodology of counting individual tenant farmers instead of just the farm owner—but they do signal decentralization, and the corresponding increase in edge effects also created an environmental change that benefited quail and other wildlife.

The amount of old field acreage also increased in the years after the Civil War. A mixture of fluctuating cotton prices, an unstable labor supply, and soil erosion and the corresponding decrease in soil fertility gave many planters little choice but to withdraw land from cultivation at particular moments. The 1880 census reported 17,204 acres of old field land in Leon County, and 20,833 acres in Thomas County.[76] Moreover, in the same year the Florida Geological Survey estimated that only half of the Red Hills region was in cultivation, the rest being in either woodland or old field.[77] After many years of cultivation, an abandoned field converts in the first ten to fifteen years to successional grasses like broom sedge, bull grass, and Florida beggarweed, depending on the available seed stock, all of which provides prime food or cover for quail. Quail, though, despite being known as a farm bird, did not confine themselves to field edges and old fields. The longleaf pine woodlands, as maintained in the Red Hills, also provided prime habitat. The yearly burning of the forests kept the understory low and free of any midstory plants that would choke out the birds' food supply, thus mimicking the early successional stages of edges and old fields.

Early northern visitors hunted quail in much the same fashion as white and black southerners. With guidance from locals, they simply took gun and dog to the nearest fields or forest and began the hunt. In the early years, landowners saw little incentive to charge for hunting rights, and the traditional informality of property boundaries persisted. Visitors easily gained permission to hunt, or more often, did not have to ask. Writer and sportsman Charles Hallock found Leon County to be especially rich for the hunter:

> To the sportsman, the prospect is admirable. In every direction, for miles from the town, are wide fields, which swarm with quail. A fair day's shooting—allowing the sportsman to take his breakfast at a reasonable hour, and start leisurely, returning for supper at dark—for a good shot, and with a good dog, is not less than from sixty to one hundred and forty birds. The coveys are all large, and often two or more are found in one field. . . . There is abundance of accommodation in the city, and the young gentlemen take pleasure in giving the sportsman all necessary information and assistance.[78]

By the 1880s locals could expect during the winter months to see northern tourists knocking about both town and countryside with gun in hand. Dr. John T. Metcalf had full access to Thomasville and surroundings, and seemed perfectly at home: "Within sight of my bedroom window I have made a bag of ten quail and eleven snipe. One can do it now, if permission be given by his honor the mayor, my great friend and crony, who lets me shoot anywhere within the city limits."[79] The casual hunting atmosphere that Metcalf described would not last. Within a few short years, northern quail hunters developed elaborate rituals to distinguish themselves from local hunters.

The movement of hunters from field edge to open woodland was central to the quail hunting experience in the Red Hills. To facilitate movement, northerners modified the road-ready horse and carriage to ride more fluidly over the rough ground of the woods. The openness of the longleaf pine woodlands already suggested a certain ease of passage, so instead of following dogs from covey to covey entirely on foot, northerners developed a hunt based out of the horse (or sometimes mule) and carriage, eventually arriving at a design in the 1880s known as the Thomasville hunting wagon. As one participant remembered, the hunting wagons' "chief characteristic was their rugged simplicity; strongly but lightly built of stout wood with four large, wooden, iron-rimmed wheels, high clearance. . . . Each carried two leather-covered seats, stiff and not too luxurious, and a dog crate of wire in the rear."[80] Hunters, usually driven by African American drivers and followed by dog handlers on horseback, rode to their hunting grounds on main roads, then veered off into the fields and forest to begin the hunt. Another observer remembered that upon reaching the hunting grounds, the dog handlers would release the dogs—either pointers or setters—and "the shooters ride in these comfortable wagons following in a very leisurely manner through the open pine woods and fields after the dogs and the workers on horseback. When a covey is pointed, those whose turn it is to shoot get out and approach for the shot."[81] Hunters alternated in twos throughout the morning, until servants brought out a lavish picnic lunch, and the hunt continued in the afternoon until they covered a previously set amount of ground. By the 1890s Thomasville became known as the home of this highly stylized, and fashionable, quail hunt. Like the conspicuous leisurely drive, the pageantry of the Thomasville hunt made clear it was a leisure activity of the upper classes.

Local landowners quickly seized on the northerners' enthusiasm for quail. By the turn of the century the leasing of hunting rights added to the coffers

Figure 4. Hunting wagon, undated photo. Quail hunters rode to the woods in style in what came to be known as the Thomasville Hunting Wagon. This early model lacks the gun and dog boxes that became common features of later models. Dogs, dog handlers, and additional servants for lunch afield would not be far behind this wagon. Courtesy of the Thomas County Historical Society.

of landowners large and small, and it was often a precursor to title exchange as well. The acquaintance of John Metcalf's "friend and crony," Mayor H. W. Hopkins, was not yet required for those visitors who wished to hunt quail, but it was fast becoming beneficial indeed. Hopkins almost invariably acted as mediator between landowners and sportsmen. From the 1880s until the late 1920s, no one possessed more control over the Red Hills countryside than Hopkins. Known as "Willie" to kith and kin, and "Judge" to the rest, Hopkins filled many roles in Thomasville: mayor, judge, state legislator, lawyer, sportsman, ombudsman, and, perhaps most importantly, real estate agent. Toward the end of his career he was able to boast that he had "located nearly every Sportsman owning preserves in this neighborhood."[82] The nephew of Dr. Thomas Hopkins, he came from a strong Thomasville family lineage, and early on he recognized the transformative possibilities in the flow of northern capital to the Red Hills. In his various roles, he knew which landowners were ready to sell and which northern visitors were eager to buy, and he often speculated in cheap agricultural land before selling to northerners. In the process he gained a great deal of wealth and influence, and he maintained a vigilant eye over the comings and goings of the Red Hills.[83]

Once landowners recognized the interest in quail as a money-making opportunity, there were three primary ways for northerners to gain access to quail land. For the less serious sportsman, most resort hotels leased land and provided guide services. As late as 1914 Tallahassee's Leon Hotel continued to lease thirteen thousand acres for any guests wanting to hunt quail. Both the Piney Woods and Mitchell House leased land for many years as well.[84] A more exclusive group of northerners—those who returned year after year—began to assert far more influence on the area's landed resources than the hotels and their one-time guests ever could; and they commenced to lease land for the season as individuals or in partnership. Or as was increasingly the case, they purchased land outright.

The major push to buy and sell land was concomitant with Thomasville's decline as a major health resort. By 1900 the many visitors of previous years began to bypass the uplands of the Red Hills for Florida's booming coastal resorts. Henry Flagler's Palm Beach and Henry Plant's Tampa Bay were in particular demand, but the infrastructure they built also led the way for a spate of smaller resorts up and down both the Atlantic and Gulf coasts.[85] The Red Hills no longer welcomed a large number of visitors, but in the minds of many, quality spoke more loudly. One visitor from Savannah observed that "The crowd goes to the east coast of Florida. . . . But while the rabble may have gone after new sights and sensations, the best element . . . has anchored in Thomasville and built beautiful winter quarters."[86] Writers of the promotional literature seized on the theme of genteel exclusivity to differentiate Thomasville from the coast. One pamphlet proclaimed, "Thomasville's first appeal is to the wealthy man who desires to winter in the South's congenial climate yet would be away from the ostentatious hilarity of the fashionable set that feels it incumbent upon itself to grace the sands of the seaside resorts further south. Thomasville is quite as fashionable, but in a more sedate manner."[87] In its decline as a tourist town, Thomasville repositioned itself as an enclave of taste and grace.

While other tourist areas became more commercialized, and new middle-class consumers increasingly looked to fulfill their wanderlust, industry barons like Mark and Mel Hanna, John D. Archbold, and Oliver Hazard Payne sought to cordon themselves off from the "ostentatious hilarity."[88] With the help of locals like H. W. Hopkins, they did so in the Red Hills by purchasing struggling farms and plantations, and piecing them together into their own private sanctuaries, causing a demographic shift in landownership that had lasting effects on the Red Hills environment and populace. For some locals in the right position, the shift meant increased wealth and influence; for small landowners, it meant selling out and moving either to town or another part of the countryside; for nonlandowning tenants, it meant different bosses in the same system; and for the new landowners themselves, the creation of these sanctuaries meant the preservation of an upper-class aesthetic thought to be lost in a quickly commercializing world.

On its surface, the sale of a plantation to a northerner constituted the removal of that piece of land from economic production. Buyers such as John Archbold had no desire to become cotton planters. A closer look, however, reveals that along with the land came labor, and new owners were not inclined to remove tenants from the land. In fact, the presence of a labor force, as a potential source of income, was attractive to prospective buyers. Tenant contracts carried over when new landowners took control, and in many cases they helped to seal the transaction. There were many variations on the tenant contract, but most of them worked out well for the landowner. This was an appealing selling point that H. W. Hopkins used repeatedly. In the early 1910s, Hopkins informed Sydney E. Hutchinson, a Philadelphia industrialist and financier, of a 1,340-acre parcel of land as a good "investment on account of the interest it would be made to pay. . . . [I] am enclosing the statement of rents for 1910. You will see that the property is paying over 7% net."[89] Opting for a 7 percent annual return on investment was not the worst financial decision a businessman could make, but with the amount of return tied to the price of cotton, neither was it the most stable. By the mid-1910s cotton prices fluctuated wildly due to overproduction and unstable markets in wartime Europe.[90] Returns on investment fluctuated accordingly. Hopkins, though, had cause for optimism in discussing a potential sale to E. B. Eppes: "The rent role of the Diamond Plantation amounts to 21,665 pounds of cotton, which at ten cents would amount to about three percent interest on a price of twenty dollars per acre, but I think the rents could easily be increased, as the late Mr. Diamond was very easy on his tenants and let them off much lighter on their rents than the average plantation owner. In this way many of the renters cultivate more land than they are strictly entitled to, and could be made to pay a larger rent by careful management."[91] As cotton prices remained low throughout the 1910s and 1920s, many quail plantation owners heeded Hopkins's advice to tighten control of rents and production.

At the same time, though, it became increasingly clear that a quail plantation was anything but a lucrative enterprise.

When the first northerners began buying land in the Red Hills, land prices were directly tied to commodity prices, but that changed as more visitors became interested in establishing retreats in the Red Hills. By 1920, Hopkins no longer classified the Red Hills as agricultural land, even though a great deal of tenant agriculture remained. As he explained it to Chicagoan Charles H. Thorne, "Cotton at 40 cents and corn at $2.00 per bushel was the prime cause" of the early century increase in land prices. "Recently," he continued, "cotton has dropped below 20 and local corn is sold from wagons on our streets at 75 and 80 cents per bushel. Owing to these conditions and the banks having all tightened up on loans, the real estate is at a standstill in farm land business and my lists are crowded with property with very few purchasers. This is the condition that applies generally to agricultural lands." On first glance, this was bad news for Thorne, who bought his 3,828 acres in Leon County for $44,477 and wished to sell for a profit. But Hopkins reassured him that there was a large and flourishing market for quail land quite separate and distinct from the market for agricultural land. The decline in agricultural land, Hopkins wrote, "does not necessarily apply to your property, and I would not think of offering it as farm land. In my opinion, it is one of the most desirable holdings for sportsman able to own it that there is in the State of Florida. . . . I would say that a price for the land of around $75,000 would be about right."[92] Hopkins obviously did not have a buyer from the South in mind. This new market in quail land was relatively immune to the economic fluctuations of the southern economy, and while the high commodity prices early in the century were welcome perquisites, the lows did little to curtail northern buying.

In fact, low cotton prices spurred a proliferation of northern-owned quail preserves. As Hopkins's comments to Thorne indicate, there was no shortage of available property; the trick was to piece small farms together to create tracts of land large enough to satisfy northern tastes. Quail preserves were to be estates, with all the advantages of exclusivity and privacy that a large spread of land affords. The buying typically began with the purchase of a larger plantation. For example, when local planter John Linton left five heirs to settle his estate, Cleveland businessman Jeptha H. Wade stepped in to buy the Linton plantation, thus creating a base from which to begin what the *Thomasville Weekly Times-Enterprise* called his "campaign of prospecting." Editor S. R. Blanton reported that Wade's "land owning appetite, once aroused within him was not easily put aside, and with more than three thousand acres of land already credited to him, Mr. Wade still purchases." Blanton continued, "The way this land owning appetite gnaws at the wealthy northerner is peculiar. It is finding illustration around Thomasville nearly every week. They never get quite enough."[93] Indeed, Wade's buying spree was not unique. In years of low commodity prices, it made little economic sense for local landowners to reject inflated offers from northerners in search of quail land.

In one respect, the nature of these purchases created its own demand. As men like Wade expanded their holdings, the amount of land available for lease dwindled, causing lease disputes between northerners and even more inflated offers to buy. Of the heavily wooded land in southern Thomas and Grady Counties, Lewis Thompson—a Standard Oil heir—thought in 1916 that it would not "be a great while before we will not be able to lease lands for shooting."[94] Until that point, the leasing of locally owned land in that area was largely controlled by Thompson and his like-minded neighbors—Archbold, Charles Chapin, and Thomas Chubb—and facilitated by Hopkins. These neighbors, though, had been in the Red Hills for about twenty years by this point and worked out standing relationships with local landowners, as well as with one another, regarding who purchased leases on particular properties.

When new northerners entered the picture they sometimes upset the equilibrium of "the neighborhood." Sydney Hutchinson, for example, began purchasing a few small tracts in Grady and Leon counties in 1910, and eventually amassed over twenty thousand acres to form Iamonia Plantation.[95] In the process of expanding his holdings, Hutchinson and his partner, Gerald Livingston, also secured hunting leases on more land—land that Thompson had previously leased. The going rate—more accurately called the fixed rate—for leased land was five cents per acre, but rumors abounded that Hutchinson and Livingston offered up to ten cents, thus causing locals to cancel their arrangements with Thompson and others. Initially, Thompson reflected that Hutchinson did not seem like "that particular kind of S.O.B.," but Hopkins was less sanguine.[96] He called Hutchinson's actions a "cowardly stab in the back," and informed him they "had never had any trouble of this kind until your appearance upon the scene."[97] After much bickering and misunderstanding, Hutchinson ostensibly worked out his differences with Hopkins and Thompson, but only two years later, in 1914, Thompson retaliated. According to Hutchinson, he "took another dig at me by buying another place over which I had shot for years."[98] The competition to lease land led sportsmen like Thompson to bypass the lease process altogether. Some local landowners suddenly began fielding offers of twenty to thirty dollars per acre or more, leaving them little choice but to sell out. As northerners began purchasing previously leased land outright, they secured control over an expanding portion of the Red Hills countryside.

In the first two decades of the twentieth century, land titles transferred from residents to nonresidents at a rate no one could have predicted. Gradually, the quail plantations multiplied and expanded, eventually merging to form what was practically a singular unit. In 1919 alone, Lewis Thompson purchased forty-four different tracts of land ranging from 7 to 680 acres to expand his Sunny Hill Plantation in Leon County, which eventually encompassed about 20,000 acres.[99] Thompson became one of the more influential preserve owners, but his name is only one among a long and illustrious list: John D. Archbold, one of the original Standard Oil trustees, amassed over ten thousand acres to form Chinquapin Plantation; Cleveland's Hanna family pieced together acreage running into the tens of thousands to form Elsoma, Melrose, Pebble Hill, and Inwood estates; Philadelphia industrialist Clement A. Griscom bought over ten thousand Leon County acres to create Horseshoe Plantation; New York banker and bakery heir, Udo Fleischmann, cobbled together sixteen thousand acres in Leon County for Welaunee Plantation; New Jersey senator Walter E. Edge and another Standard Oil scion, Walter C. Teagle, eventually purchased more than eighteen thousand Jefferson County acres and named it the Norias Club.[100] This is only a partial list. Far from a mindless buying spree, the intent was to ensure the utmost privacy.

Sharing one's borders with those of another quail plantation meant a solidarity of purpose: the de facto enclosure of the open range. No longer could neighbors cross ownership boundaries without a thought; there were few landowning neighbors left. Of this enclosure, Hopkins wrote to one potential buyer, "You will note from the map that the tracts of land I mention are surrounded by other large game preserves. This ensures protection from poachers, as all of these parties are wealthy and influential, and combine their efforts to prevent poachers from interfering with their rights and privileges. This gives quite an advantage, as compared with a community where sentiment in favor of poaching prevails."[101] Hopkins's implication was that local landowners would feel free to cross property lines as they had in the past, and the easiest way to close the open range was to have like-minded neighbors who respected rigid ownership boundaries. What was once a community right to access was now poaching, and unimproved land in the Red Hills gradually fell under the individual's right of enclosure.

Enclosure of the range was not the exclusive province of northern quail hunters, of course. Southern planters were also engaged in the effort. The right of landowners to control access to unimproved land was a relatively new phenomenon in the South; until the Civil War, they were much more concerned about their right to control slave property.[102] The southern range was open, allowing small herders and farmers to graze livestock, hunt, and fish on any unimproved land.[103] As their source of wealth transferred to land, large planters pushed to close off the range through fence and stock laws, as well as by strict enforcement of trespass laws. When northern quail hunters entered the picture, though, this transformation was hardly complete. Some laws restricting hunting were on the books—Thomas County passed a law in 1876 making it "unlawful for any person . . . to shoot, snare, trap, or kill in any manner, any wild turkeys or partridges" during the mating and nesting seasons, between March 1 and October 15.[104] But during the hunting season, residents of both town and country continued to visit old hunting grounds, and when northerners began leasing and buying land, predictable confrontations ensued. Some northerners were nonplussed by the South's custom of the open range. Ned Crozer, who held a lease in Leon County, complained that "it looks to me the way things stand now as if any one could shoot on our lands;" he was "almost willing to do anything to protect myself against trespassers, as most of them would be only too glad to get a chance to bush the quail in our country."[105] Crozer's concern was less one of wildlife conservation than one of possession. Quail, unlike most other game animals, occupy a small range and spend most of their life within a quarter mile from where they are born. Crozer's mind worked more like that of a British landowner, whose rights of ownership included control of all that was on the land; and since quail were not likely to leave his land, he held possession over them. To Crozer it was only logical that quail on his land belonged to him, and that his neighbors had no right to encroach on his property, fixed or movable.[106]

Quail preserve owners, like local planters, occasionally looked to the law to resolve their problems, but successful enclosure came more often by simply continuing to buy land, thus dispensing with neighbors and reorienting local custom. The open range was, after all, borne of tradition and custom, and could be repealed in the same way. To paraphrase Marc Bloch on enclosure in the French context, the violation of custom itself became a tradition.[107] The legal right to post land under fee simple ownership was always present for landowners, but because of long-established community custom, it was very difficult for southern planters to keep others off their land.[108] Once northerners possessed so much land, they erected fences around property boundaries, posted land, and dealt with violators on an individual basis. Before Udo Fleishmann began buying land, he leased from no fewer than ten landowners in 1910, and announced in the *Tallahassee Weekly True Democrat* that these "lands are posted and all persons are warned not to hunt on said lands as they will be prosecuted to the extent of the law."[109] The law did not extend much further than a small fine, but Fleischmann's shot across the bow served notice that local custom was quickly shifting.

Even when preserve owners posted their land under fee simple rights, monitoring and enforcement proved difficult, especially when few authorized hunters were in the woods during the out-of-season months. There was little public protest about the loss of hunting rights in the Red Hills, but many did resort to subterfuge. Those areas on the edges of the Red Hills seemed especially vulnerable to poachers slipping in and out of the heavily wooded areas without detection. Landowners instructed overseers and tenants to keep a close eye out for any suspicious looking characters. Hopkins monitored many estates when preserve owners were not in residence and contracted with local informants to report any possible poaching. When one unknown informant reported that poaching was "getting to be general again" on parts of Susina Plantation during the spring and summer of 1912, Hopkins responded by hiring a full-time game warden to patrol the land and turn in poachers to law enforcement for hunting out of season.[110] The owner of Mistletoe Plantation remembered being "worried all the time by poachers." Despite building "5026 miles of fence around the place" between 1914 and 1918, "Cairo men came down in the middle of the night and camped to be ready to shoot at light."[111] Indeed, "Cairo men" seemed to act as a synonym for "poacher." The seat of the newly formed Grady County, the town of Cairo was situated in an extensively farmed section east of the Ochlocknee River, where deer hunting was notoriously poor.[112] Many of its residents became so adept at easing across the river into the heavily wooded section to the east that most preserve owners did not want the publicly accessible stream as their boundary. In shoring up his borders, John Archbold figured "there is something like 275 acres lying North and West of the river that I don't own," and inquired of Hopkins, "Do you suppose it would be possible for me to buy the balance of this land now at a reasonable figure? I would be glad to do so as I don't like the river as a boundary."[113] "Cairo men" continued to spill into the Red Hills, however. Years later, Herbert Stoddard complained to Aldo Leopold about poachers on his Grady County property, saying "our poaching situation is very bad all the way around. . . . Certain bad actors are shooting deer day and night."[114]

Though there were few public protests about the loss of hunting rights, many farmers did lament nonresidential ownership of good agricultural land. Public calls to curtail northerners' purchase of land occasionally appeared in the Tallahassee newspapers. The *Weekly True Democrat* editorialized in 1914 that the large preserves prevent "the prosperity we are so anxious to see. Small farms are the true source of dependence, and the policy that prevents an increase of population is wrong and damaging."[115] Rudolph Herold, a native of Switzerland who arrived in Leon County by way of Iowa, in 1913 could not "see that farming conditions are any better than they were when I moved here" seventeen years earlier. He reserved his true ire for Tallahassee's merchants who buy "everything we grow way below market prices," and added, "What business has any white farmer with one grain of sense left to remain in this sickly malaria country?" This was, indeed, a far cry from the flowery prose northern visitors had penned about the area just a few years earlier. Herold concluded sardonically, "I think we all had better turn it over to the colored race, or else sell it to northern sportsmen, so they can post it, build high fences around it and raise snakes and birds."[116] Farmers like Herold found no utility in birds and snakes; they were out to make a living from the land, not cordon it off as a landscape of leisure.

This divergence between what local farmers and preserve owners expected from the land was real, but such public opposition also reveals a great deal of white apprehension over black tenants having access to so much land. In many ways Herold's conclusion reflected reality: local whites had, in fact, sold to northern sportsmen, and at the same time turned the land over to African American tenants. The problem from the local white perspective was that tenants continued renting land and farming cotton with northerners and not local planters. Many local whites felt the trend toward northern ownership a colossal waste of prime resources. Another 1920 editorial in the *Tallahassee Daily Democrat* proclaimed "Untilled Lands a Leach," and recommended "summary action against the individuals or corporation who buys up large tracts of rich farm lands and refuses to cultivate them or sell them to people who will."[117] Such criticisms usually neglected to mention that a great many people were still turning the soil and making a living on the quail preserves.

Despite the protestations, northerners continued to expand their holdings, closing off the range to outsiders as they grew. For insiders, however, the preserves became a type of private commons. Those tenants who remained on the quail preserves and knew the land best generally had free range to hunt (sometimes even quail), fish, and, in the early years, graze livestock throughout the preserve environment. There were limitations, especially as preserve management became more regimented in the 1920s and 1930s, but both African American and white tenants and employees continued to use the land much as they had for many years previously. Legally, the common fixed-rate rent—as opposed to farming on shares—gave tenants temporary possession of their parcel, and, moreover, there is little to indicate a reduction in traditional access to former common areas.[118] Henry Beadel, owner of Tall Timbers Plantation, continually remarked with little apparent surprise or judgment in his diary of tenants in "a blind or boat shooting at everything that came along," or of their roaming "droves of pigs" interrupting a quail hunt.[119] The overwhelming number of tenants who remained on the preserves after exchange of title is evidence enough of their access. Henry Vickers, a tenant born on Tall Timbers, estimated that at least two hundred people lived on the 2,500-acre property in the early 1900s.[120] At least thirty-nine tenants worked on James Mason's Susina Plantation in 1909, not to mention their families.[121] And about four hundred people lived on Horseshoe Plantation as late as 1930.[122] Tenant populations dwindled on the quail preserves through the twentieth century—as they did in all of the South's plantation regions—but until the 1930s, and later in many cases, tenants and their families continued to use the preserve environments as a type of private commons on which they scratched out a living.

For many northerners, the decision to keep tenants on the land was as much about meaning and power as it was economics. Rents could supplement the cost of running a preserve—and sometimes provide a profit for landowners—and tenants sometimes fulfilled the role of sentries on the lookout for poachers. But equally important to landowners was maintaining the social relations found embedded in what appeared to be a picturesque nonindustrial landscape, one that was uniquely southern. One of the early selling points by health-trade boosters was the "quaintness" of the Red Hills' laboring class. Photographs of lounging African American children with captions like "Who Cares?" appeared in numerous tourist pamphlets, as did scenes of adult tenants and their oxcarts captioned with "Country Come to Town."[123] One northerner observed that some "natives can move wonderfully slow, if the eye of the boss is not turned their way," and another considered "the negroes . . . a source of infinite amusement."[124] After spending six years in the region, and coming to know the preserve owners well, Herbert Stoddard noted in 1931 that many preserve owners had "a well-founded affection for these people, with their quaint speech and many sterling qualities."[125] To the urbanite fleeing the hustle of the city, such scenes conveyed the social dynamics and hierarchies that were long part of the Old South myth—"natural" images of docile black laborers and genteel white landlords. These northern captains of industry adopted a sense of themselves as paternal caretakers not unlike the Southern planters of old, and seeing African Americans as picturesque—suspending them into a desired aesthetic—successfully obscured the South's oppressive labor system.[126] Moreover, in naturalizing black laborers as part of the landscape, in buying up the countryside and growing their estates to previously unseen proportions, and in having little preoccupation with balancing the plantation books, northerners in the Red Hills actually accomplished more toward reaching an Old South ideal than southerners ever had.[127] They were after an aesthetic of pre-industrial production not unlike that of serfdom, and this iteration had a distinctly southern cast.

Many new owners went about constructing picturesque spaces quite intentionally, which in some cases meant refining the very environments they found so attractive in the first place. Upon purchasing an estate, most owners constructed lavish homes—or renovated older plantation houses—and installed elaborate ornamental gardens. They also devoted a great deal of design attention to their land beyond the house grounds. Jeptha Wade, for example, hired the renowned landscape architect Warren H. Manning—a protégé of Frederick Law Olmsted—to design the house grounds at Mill Pond Plantation. Beyond the grounds, Manning and his team were especially impressed with how Wade's longleaf pine woodlands meshed so well with their own aesthetic ideas. Manning's landscape architecture stressed the "wild garden," wherein "the Landscaper recognizes, first, the beauty of existing conditions and develops this beauty to the minutest detail . . . instead of by destroying all natural ground cover vegetation or modifying the contour, character, and water context of existing soil."[128] Manning's goal was to leave human design as invisible as possible in his landscapes, and on Wade's property he was especially pleased with "the richness of the native flora . . . in which the number of evergreen species that will count most effectively in [the] winter landscape is large." Such richness helped him to achieve "the coveted evergreen effect in winter without artificial planting." The woodlands already reflected Manning's desired aesthetic. Open, with plenty of sunlight and interesting views, it was park-*like*, the perfect "wild garden." Manning advised that the longleaf forest was "to be let alone at present."[129]

But Manning and the preserve owners did not realize that to "let alone" the longleaf pine–savannah woodlands was to change its very nature. Its complex ecology, as well as its aesthetic, developed in concert with human work. This natural landscape and its most appreciated inhabitant, the bobwhite quail, could not sustain its desired conditions without many of the cultural traditions of southern rural life, or at least a conscious impersonation of them. As the Red Hills quail preserves matured into a more stable entity in the early 1920s, the aesthetic remained on most estates, but the bobwhite quail population began to decline. Preserve owners began noticing the decline in the late 1910s and early 1920s and had little idea of the cause. The most frequently cited cause by those with the virtue of hindsight has been the exclusion of fire in compliance with federal and state campaigns that preached its evils, but only on a few preserves, such as Jeptha Wade's, did landowners and tenants actually stop burning.[130] On the contrary, there is abundant documentation that annual burning continued throughout the Red Hills even during the most intense anti-fire campaigns. The cause of the quail decline was more likely a combination of altered agricultural practices, hunting pressure, and natural fluctuations in quail breeding cycles and predator-prey ratios. But again, there is little evidence to suggest agricultural practices changed that much until the 1930s, and an attempt to gauge cycles and ratios would be fruitless in this study. So on the basis of historical evidence it is difficult to say for certain why the quail population declined—nor is it really necessary to our larger story. The preserve owners' reaction to the decline, however, is significant. They turned to the U.S. Bureau of Biological Survey, who would send Herbert Stoddard, a field agent who had an immeasurable impact on conservation and ecological understanding, not only in the Red Hills but also in the region and nation. Herbert Stoddard's move to the Red Hills would make the quail preserves much more than a landscape on which to create an aesthetic. They would become an experimental laboratory, a breeding ground for the new profession of wildlife management, and a central source of dissent against the growing influence of modern agricultural and forestry interests in the South.

By the 1920s the Red Hills was set to develop and profit from a new identity. No longer was it a place of the failed southern plantation, trying to hang on to the Old South. It was now known as the domain of the wealthiest, most refined class in America. What began with Gilded Age and Progressive-era concerns over health resulted in American nobility taking over this particular corner of the southern countryside. But beneath the veneer of the tasteful, northern-owned quail preserve lay the same structures of power that propped up southern-owned plantations. Sharecropping and tenantry continued relatively unchanged, but there were some very important differences in perspective between the old southern guard and the new northern owner. The latter came to the woods to escape the processes of production. They easily left the factory behind, but the production of the countryside was something else. Their travels by rail exposed them to the ravages of the timber industry, and in purchasing as much land as they could, the northerners successfully distanced themselves from its advance. The farm, on the other hand, was ever present, but in the longleaf pine woodlands, its nonindustrial aesthetic made it seem to northerners like a natural part of the landscape. Farm patches scattered throughout large blocks of woodland created a mosaic effect that not only seemed natural but also created the desired aesthetic and environmental conditions northerners wanted in the countryside—and critically, it created a lot of wildlife. In naturalizing tenant agriculture as part of the landscape, the new landowners also naturalized the black tenants themselves, thus successfully obscuring the oppressive inequalities of tenantry. Raymond Williams has shown the landscape of the English manor to be built on the backs of the laboring classes, and that what became picturesque borrowed many details from their working landscapes. The quail preserves of the Red Hills came about in much the same way, but whereas Williams's English countryside was by most accounts a constructed landscape with little resemblance to that environment's historical range of variability, the aesthetic of this working landscape in the Red Hills integrated well with the disturbance-dependent longleaf pine woodland environment.[131]

This integration of nature and culture was happenstance, an unconscious result of ecology, modes of production, and ideology. It was what some scholars have called "second nature," a mix of environmental and cultural processes neither wholly natural nor entirely artificial.[132] But as the region's political economy continued to change throughout the first half of the twentieth century, and the norms of production strayed further from those of ecology, the continued maintenance of the longleaf pine–savannah environment required a shift in ideology. This new ideology, based in the biological sciences, would result in an intentional effort of land management and conservation. Herbert Stoddard's arrival in the Red Hills began the process.

CHAPTER 2

The Development of an Expert

Herbert Stoddard's journey south in early February 1924 must have been a little nostalgic. He had spent eight childhood years in the longleaf pine forests of central Florida, trapping a variety of mammals and reptiles, running with cattle herders, amassing a collection of wild pets, and generally running roughshod over the forest. This was his first visit to the southeast in twenty-four years, and this time he came in a far different capacity. He was now a professional ornithologist and government agent, and the work he carried out would not only change land management in the Red Hills, it would have a profound effect on conservation in the southeast and the entire nation.

As an exposition on how Herbert Stoddard, a high school dropout from a working-class family, negotiated the highly stratified, expert-driven, newly professional world of the natural resource sciences and came to fill the role of scientific expert, this chapter attempts to penetrate below the administrative surface of the making of modern conservation. Stoddard's biography not only helps us to understand the development of the modern conservation movement, and how that movement played out both regionally and locally, but it also reveals the still inchoate organization of government conservation and its tenuous relationship with the biological sciences at the end of the Progressive era. When Stoddard came to the Red Hills in 1924, there was still not much of a formalized route to conservation or scientific work, nor was there a codified method for carrying out fieldwork, especially with the study of wildlife and its habitat. Such organizational flexibility, along with his lack of formal training, led Stoddard to devise an approach that blended scientific and local knowledge and had an immense influence over the emerging field of wildlife management. How he arrived at such an influential blend can be found in his early years. Stoddard's circuitous route to south Georgia—his background as a curious kid, his informal training as an apprentice taxidermist, and his coming of age as a professional ornithologist and field worker in natural history museums—is a crucial part of the Red Hills story. He was, after all, thirty-five years old when he became an agent of the Biological Survey. To gloss over his early years would not only slight more than a third of his life; it would also obscure how his early career suggested, even embodied, important developments in the transition in science and conservation from the Progressive era to the interwar years.

The line between amateur and professional scientists remained fuzzy until well into the twentieth century. The road to becoming a scientific specialist included many detours into local communities, commercial enterprise, corporate institutes, and public and private natural history museums, all of which helped to shape the type of science and conservation experts in the field would develop. Field scientists like Stoddard also had a great deal of contact with working landscapes and local knowledge, and his training came in these local contexts, not at a university. Specialized training in the biological sciences began to require higher education during this time, but people like Stoddard continued to be viable scientific actors well into the twentieth century.[1] Stoddard struggled to complete primary training and never made it to secondary school. His professional training came through what most closely resembled a series of apprenticeships, instead of the specialized rigor of the university. But he became fluent in the science of systematic biology, or taxonomy, the bread and butter of the biological field sciences around the turn of the twentieth century.

Though he never received a formal education, Stoddard's early professional training was very much in line with that of his peers. Taxonomy was still the baseline of the biological sciences, and natural history workers like Stoddard took to the field to catalog and collect as many species as possible around the turn of the twentieth century. This "age of the survey" took place when previously inaccessible places opened up, allowing for government bureaus and natural history museums to conduct surveys both intensive and extensive of a bioregion's flora and fauna.[2] The timing of Stoddard's professional ascendancy was critical. Not only was he one of the last to reach the status of scientific professional with no formal training; he also spent his early years near some of America's most important inner frontiers. If the Gilded Age and Progressive era embodied "a search for order," as historian Robert Wiebe has argued, when "a nation of island communities" was breaking down under a web of transportation, business, and professional networks, there remained wild spaces in between.[3] These were places where natural communities remained intact for a professional class of scientific workers like Stoddard to explore and catalog. In the process, they constructed a taxonomic order through which to understand nature.

The special nature of ornithology was important to Stoddard's development. His expertise in ornithology was, after all, what brought him to the Red Hills, and the result of his quail study was the birth of yet another specialty, wildlife management. But ornithology was a weird bird, one of the few remaining biological disciplines in the early twentieth century that retained an egalitarian spirit and resisted academic professionalization. Even while Cornell University, the University of California at Berkeley, and the University of Michigan established graduate programs in ornithology in the 1910s, scientific ornithologists continued to rely on "self-education and apprenticeships to familiarize themselves with the set of practices and the body of knowledge associated with their field," according to historian Mark Barrow.[4] The broader traditions of natural history held their ground in ornithology, allowing professionals and amateurs alike to venture into nature making bird lists, chronicling bird behavior and habitat, and comparing notes. Not only that, but bird species were among the best documented animals in the taxonomic array by the 1920s, which led ornithologists to look for new avenues of discovery. The new field of ecology—fitting the taxonomic parts together into a functional whole—began to show a great deal of promise during these years, and its perspective would be vital when Stoddard began to work on the life history of the bobwhite quail. Rather than studying bird groups in isolation, he and other ornithologists began to see birds as they related to each other, other animals, and their habitat. Stoddard's career path represents a bridge between taxonomy and ecology, and an understanding of his background as a museum taxidermist, biological field worker, and ornithologist is an essential first step in understanding the brand of conservation science he created upon arrival in the Red Hills.[5]

Herbert Stoddard's appointment to the Bureau of Biological Survey in 1924 marked the culmination of an informal training in natural history that he began as a youngster in the piney woods of central Florida. Stoddard was born and lived the first few years of his life in the upper Midwest, a region to which he would return as an adolescent, but it was during the seven years that he spent in Florida that he came of age. When the Stoddard family arrived in the town of Chuluota in 1893, they must have thought the train line from Chicago doubled as a time portal. At the train station in Chicago, crowds flooded in to view the World's Columbian Exposition and the progress and promise of modern America. For the Stoddards, Chicago was simply the main point of embarkation. They came from their home in Rockford, Illinois, and had little time for the Columbian Exposition. They did not see the White City or the Midway Plaisance; nor did they hear Frederick Jackson Turner announce the closing of the American frontier. They were on their way out, joining a transformative wave of immigration to Florida, where land was still cheap and plentiful. If they had heard Turner's proclamation at the Columbian Exposition, they would not have believed him upon arriving in Florida. Unlike the health seekers that flooded the lavish coastal resorts, Stoddard's family looked inland to capitalize on the bourgeoning citrus industry, and while there they witnessed a peninsular frontier that seemed anything but closed.[6]

Most of what we know about Stoddard's early years in Florida comes from his autobiographical recollections fifty years removed from the time and place, which read not unlike the post–Civil War travel writers who promoted the state. His years of experience as a wildlife biologist, forester, and ornithologist lend the narrative considerable credibility, though it is sometimes peppered with a healthy dose of nostalgia. He witnessed much industrial logging after his return South in 1924 and lamented deeply the loss of the region's longleaf forests. During Stoddard's youth in the 1890s much of Florida's interior remained untouched by industrial timber companies. For a kid with a penchant for the outdoors, the region's pine and cypress forests were a dreamland: "No one seeing the cut-over, devastated forests of Florida today can possibly imagine the beauty and grandeur of those woodlands. One could ride through them for a week on horseback, and still they stretched on and on, broken only occasionally by a settler's tiny clearing."[7] Images of nature undisturbed are everywhere in Stoddard's memoirs, but so, too, are those of working landscapes. Despite its frontier status, Florida was on the upward arc of an unprecedented land boom.

Throughout the post-Reconstruction period, as rail spurs fingered across the landscape, the state of Florida unloaded more than 10 million acres of the public domain to real estate or timber interests, and immigrants flooded into the peninsula to reap the fruits of Florida's subtropical climate. The Stoddards most likely discovered Chuluota and surroundings through one of Henry Flagler's many local land companies, and like other unsuspecting northerners, their attempt at establishing orange groves ended in failure.[8] The years in Florida were lean times for Stoddard's family. Herbert's father had died back in Rockford, and though his new stepfather "was a fine man, he never found his place in life . . . [and] had no talent for business matters or the earning of a dollar."[9] The family mostly subsisted on the life insurance settlement from his father's death, though that was little help in Florida's land-boom climate. Stoddard wrote that the "'land sharks' took our family for the same kind of ride given thousands of others in the 'gay nineties.' By the time most of our land was cleared and ready for orange groves nothing remained of their savings."[10] Adding insult to injury, Florida's freeze of 1895 wiped out the young orange groves of the entire region, and the Stoddards found themselves among other northerners "marooned penniless in a country of which they knew all too little."[11]

Regardless of his family's economic failures, the venture allowed Stoddard to explore the surrounding forests and develop a love for wildlife and wildlands. He collected snakes, alligators, and tortoises as pets, and contributed to the family's sustenance with countless hours spent in the woods hunting, trapping, and tracking an assortment of game. He even supplied a growing commercial trade in natural history by selling skinned animal hides to family connections in the Midwest.[12] In Stoddard's mind, though, his most enduring lessons came from his interaction with local cattle ranchers, who taught him the value of what he came to call "woodsmanship." He clung to these local woodsmen, absorbing all they knew about the surrounding pine forests. The many hours Stoddard spent learning the woods in Florida would prove formative in his later work.

Chuluota was situated on the eastern edge of Florida's central highlands, such as they are, in Orange County, which was the primary location of the state's burgeoning orange industry. But surrounding the highlands was the vast domain of open-range cattle herding. And cattlemen dominated the state's flatwoods interior. During Stoddard's stay in central Florida, most of the industrial timber companies remained occupied with Georgia's pine forests, and much of the former public domain, though under title, was still controlled by "cattle hunters," who retained free range over the area and usually spent most of their days rounding up cattle from the open pine forests and palmetto prairies.[13] Early on, he befriended members of a local cattle family, Gaston and Polly Ann Jacobs, and "spent more time with [them] . . . than I did at home, for the cattle work fascinated me."[14] Like most other cattle families, the Jacobses would have owned 80 to 160 acres of land, where they built a home, outbuildings, and a split-rail cowpen, and maintained garden and corn plots. Their cattle roamed up to a hundred square miles of the surrounding open range alongside the herds of other families, differentiated only by the mark of registered brands.[15] As a youngster, Stoddard's horse work was limited, but since "a boy on foot was a great help with the cattle," he spent weeks at a time on the range tending to new calves and mature cows ready for market.[16]

Stoddard recalled the Florida cattlemen with an admiration bordering on reverence: "I do not believe any part of America produced better natural woodsmen than were the cattlemen of this part of Florida. With none of the distractions of modern life, they were true 'children of nature,' with a large and much-used store of woods lore."[17] Such a sepia-toned remembrance may be too romantic; anyone involved in a market-driven business like cattle ranching was very much in tune with "modern life" at the turn of the twentieth century. But even in their engagement with modern life, their labor required intimate knowledge of nature's doings; this particular form of market-driven labor acclimated workers to their natural surroundings, rather than alienating them from it. The above passage begins to flesh out what Stoddard came to value as a wildlife biologist: his training was rooted in local knowledge and predicated on spending a great deal of time in the woods with the locals who knew them best. He came to believe that the first point of order in any unfamiliar environment was to get to know it, and the quickest way to do so was to spend time with locals who already did. These experiences in the Florida woods, then, were not merely what animated Stoddard's interest in nature and the longleaf-grassland ecosystem in particular; they formed a core part of his managerial philosophy, one that insisted there was no scholarly substitute for an intimate working knowledge of one's surrounding environment. Indeed, these "cattle hunters" were Stoddard's first model for the informed land manager he would become.

Easily the most important product of Stoddard's exposure to the Florida cattlemen with whom he worked was his nascent recognition of fire's place in southern longleaf woodlands. It was common practice for southern ranchers to "burn over and manage this vast domain as [they] saw fit," knowing as they did that the grasses, forbs, and legumes of a fire-maintained groundcover made for better grazing than the "rough" of hardwood brush found where fire did not reach.[18] The highly flammable undergrowth of wiregrass, saw palmetto, and other herbaceous groundcover was the only feed available to cattle; it was only after World War II that Florida ranchers commonly supplied their cattle with supplementary fodder. The native grasses, while not ideal nutritionally, were sufficient to maintain a market-viable stock, as long as cattlemen renewed the vegetation with annual fire.[19] After a year's growth, wiregrass became hard and dry in the winter, prompting cattle herders to burn the range to release spring growth. One cattleman remembered, "The grass—after it gets old and tough—it's not much good. There's a lot of wiregrass, and when it's fresh burned, it's real good grazing."[20] Shortly after a burn in February or March, a fresh mat of tender wiregrass—as well as young saw palmetto—came back to supply graze. The cattle herders' motivations for burning may have been self-interested, but their self-interest helped them to appreciate the historical role of fire in the region.

Like many who look back to their childhoods, Stoddard tended to wax nostalgic on occasion, but he was insistent about what those early years in the piney woods taught him. "Looking back on my early life in Florida," he wrote in his memoir, "I am convinced that no schooling or advantages could have been more valuable to me. I firmly believe that all experiences become a part of a man. Certainly my years in the southern pinelands—conditioned as they were by the forces of climate, hurricane, and fire, rooted in soils laid down under the gulf such a short time before, geologically speaking—those years were invaluable to me in my later years as ornithologist, ecologist, and wildlife researcher and manager."[21] These experiences with nature, along with the transparent manipulations of nature by the human community, were instrumental to young Stoddard's learning, and they carried over throughout his adult life. Indeed, they contributed to an informal and intimate training that would serve Stoddard well when he returned to the Southeast in the 1920s. But such local training would remain just that—local—without institutional connections that covered broader terrain. A boy with Stoddard's background required a series of fortuitous moves to locales more *au courant* for the opportunity to join a growing movement in science and conservation.

As the Stoddard family's financial situation grew worse in Florida, they decided to move back to the upper Midwest, one of the cradles of the modern conservation movement. They landed in Stoddard's birthplace, Rockford, Illinois, in 1900, where young Herb continued to cultivate an interest in natural history, a career choice with few outlets for a boy with little hope of higher education. There was, however, taxidermy, which at age eleven, Stoddard "decided upon for my lifework."[22] In the America of 1900 a boy like Stoddard had few other options in the field of natural history. There was no family money to put him on the road to become the classic amateur naturalist, and in an age of increasing specialization in the biological sciences, he had no prospect of securing a place in the academy. But this was also the age of nature study, a somewhat more egalitarian pursuit, and taxidermic display was one way for an increasingly urban population to indulge in nature.[23] As modern science rationalized nature to the point of abstraction, the nature study movement attempted to reconnect urban people with the natural world. In the urban core, taxidermy came to be one of the natural world's primary forms of representation, and demand for quality displays increased dramatically. Taxidermy was not only Stoddard's way into natural history; it also contributed to his family's financial well-being. He found and read William Temple Hornaday's *Taxidermy and Zoological Collecting* and John Rowley's *Art of Taxidermy*, then enrolled in a correspondence school and immersed himself in the craft. By early 1901 Herb already had a display of his first taxidermy efforts in a local drugstore window in Rockford. These early displays, though not yet of museum quality, included skins from both his Florida collecting and anything in Rockford and surroundings he ran across. Since "everybody seemed to want to be horrified at the Alligator, Diamond-back and Wildcat hides, there was 'standing room only' around those windows for weeks."[24]

Stoddard never had much patience for formal education and chose to leave school in 1905 at age fifteen, moving to his uncle's farm in Sauk County, Wisconsin, to work as a farmhand. He was still enamored with his previous life in the wilds of Florida, though, and "the settled farming life was of little interest to me."[25] In Prairie du Sac, Stoddard met Ed Ochsner, a local taxidermist, beekeeper, and naturalist who came to be an important influence. Over the next several years, Stoddard worked on the farm during the growing season and with Ochsner during the winter, honing his taxidermy skills as well as his trapping and collecting methods. Like the cattlemen with whom he ran in Florida, Ochsner was a prototype for the knowledgeable woodsman that Stoddard would strive to become, "a person as unusual as any of his specimens."[26] While most future scientists and naturalists spent their teenage years prepping for the academy, Stoddard was earning a living in the field, trapping, hunting, skinning, and discussing birds, mammals, and general natural history subjects with Ochsner.

There were few places better than Prairie du Sac and surroundings to immerse oneself in natural history. Today, the terrain of Sauk County is immortalized in Aldo Leopold's *A Sand County Almanac*, which was, among many other things, a paean to the particulars of place. Formed from the wreckage of glacial drift, the dominating geological feature is the Baraboo Bluffs, an oval shaped "inner frontier" of hills and quartzite outcroppings that circle through the center of Sauk County. Devil's Lake and its five-hundred-foot-tall cliffs are the gems of the bluffs, and just to the east runs the Wisconsin River, easing its way down toward the Mississippi. And, as many local place names attest, there were also vast stretches of prairie land, where the fathers of important thinkers like John Muir and Frederick Jackson Turner first set plow to dirt in the early nineteenth century. Here in this thoroughly settled country, Stoddard found nature aplenty, and through locals like Ochsner, absorbed the environmental knowledge of previous generations.[27]

Discussions with Ochsner about natural history were critical to Stoddard's field education, but perhaps just as important were Ochsner's professional connections. He had loose ties to the Milwaukee Public Museum (MPM) and the Field Museum in Chicago, and was a hunting companion of the Ringling brothers, who housed and trained their circus in nearby Baraboo, Wisconsin, during the winter. Stoddard's professional break came in 1910 on a trip with Ochsner to visit the Ringlings. They traveled to Baraboo often, and on this trip they found Alf Ringling with a dead hippopotamus on his hands. Ochsner, who was always on the lookout for potential museum specimens, decided it should go to the MPM, if they could skin and pack it for travel. The head taxidermist for the museum, George Shrosbree, came to Baraboo immediately, and Stoddard stayed on to assist in preparing the skin.

That Stoddard found his way into the museum profession through the circus is significant. Historian Janet Davis has shown that the circus, especially as exemplified by the Ringling Brothers, was a "powerful cultural icon of a new, modern nation-state," and as such was at a center of American ambivalence toward modernism at the turn of the twentieth century.[28] In its efficient use of technology, integrated business structure, large industrialized workforce, and domination of exotic wild animals, the Ringling Brothers Circus represented to spectators the most modern, and powerful, components of American civilization. At the same time, though, it revealed what modernism was on the verge of destroying. As Davis argues, "Showmen publicly mourned urban encroachment, massive immigration, and the imminent loss of the frontier; as such, they marketed the railroad circus as a place where audiences might catch a 'last glimpse' at the world's vanishing animals and preindustrial people."[29] As it was for so many Americans who encountered the Ringling Brothers' circus animals, this was Stoddard's first contact with exotic nature, and he was thrilled to be so close to it. While working inside the hippo he found the 8-gauge slug that had apparently aided in the animal's capture years before, and his imagination took over. Stalking and bringing down the creature in the wilds of Africa, with tribal Africans in tow—it was the stuff of dreams for the turn-of-the-century white American male. There was nothing in the Americas that compared with an animal such as a hippopotamus, and thanks to institutions like the circus and natural history museums, Americans interested in nature considered Africa to be the wildest of wild nature. Such firsthand contact with exotic nature set Stoddard's mind to working, as well, on the possibilities of exotic exploration.[30]

Watching Stoddard pick his way through the flesh, bones, and entrails of the hippo, George Shrosbree recognized the skills of an enthusiastic young apprentice. He offered Stoddard the job of assistant taxidermist at the MPM a few weeks later. In a professional world that was still taking shape, the informality of such an offer was not all that unusual. But it did give a budding amateur naturalist entrée into the newly developed, insular world of expert-driven natural sciences. Ringling's hippo and its transformation from commercial spectacle to scientific specimen in many ways resembles Stoddard's leap into the museum world. With Ochsner, he was an apprentice in the world of commerce, learning how to make a living on his knowledge of the natural world; with the museum, his duties would be similar, but there was a presumed elevation above commercialism. Museums had risen in lockstep with consumer society, but practitioners understood themselves to be above the fray, hovering in the realm of pure science—knowledge for knowledge's sake. Throughout the first half of the nineteenth century, many museums were for-profit and sensational, Phineus T. Barnum's American Museum being the best known. But as the century progressed, institutions like Louis Agassiz's Museum of Comparative Zoology at Harvard and the Museum of the Boston Society of Natural History became important centers of public education in the natural sciences.[31] By the turn of the twentieth century, a mosaic of publicly and privately funded museums, filled with elaborate mounted displays of both exotic and native animals, populated urban centers across the nation.

Such fluorescence reflects the growing interest in natural history among the general populace, but it was also an important step toward the development of the biological sciences as a research-based profession. The traditional developmental narrative of biology and zoology maintains a sharp linear shift from amateur pastime to academic profession, but historian Keith R. Benson has characterized it as a "gradual transformation . . . from its primary location in museum-oriented natural history," populated by both amateurs and professionals, "to its eventual setting within academic and research institutions."[32] When Stoddard began work at the MPM in 1910, universities such as Johns Hopkins, Harvard, and the University of Chicago had developed highly specialized fields of biological research and emerged as the leading producers of scientific experts.[33] Museums, on the other hand, continued on in the broader tradition of natural history, interpreting and disseminating gains in scientific knowledge for the public. Stoddard, by all accounts an amateur naturalist under Ochsner's tutelage, was now a part of the scientific fraternity, learning to become a professional. For the next fourteen years he performed his duties in the museum field, first for the Milwaukee Public Museum (1910–13), then the Field Museum of Natural History in Chicago (1913–20), and then again for the MPM (1920–24) as taxidermist, field collector, and, finally, ornithologist.

When Stoddard joined the staff of the MPM, it was among the nation's elite institutions in the study of natural history. It secured its public charter in 1882, and by 1910 was already a major interpreter of nature not only in the Midwest, but in the nation. It was the nation's first publicly chartered museum, but its ties to commercial enterprise were strong. The museum's base collection came from the Natural History Society of Wisconsin, and, like most museums during these years, it did not have a staff of field collectors or taxidermists to build display groups. Henry A. Ward's Natural Science Establishment in Rochester, New York, filled the void by mounting and installing all varieties of animal displays. Ward's was the nation's largest taxidermy and museum supply house throughout the 1870s and early 1880s, and it employed the most artistically and technically advanced taxidermists in the United States. By the end of the latter decade, however, museums took on more responsibility for collecting and preparing their own displays, and Ward's influence quickly declined.[34] By 1886 two of his finest taxidermists, William Morton Wheeler and Carl Akeley, were on staff at the MPM, setting it on the path toward developing its own style of animal display. Both became important figures in the world of science, Wheeler as a pioneer entomologist and Akeley as a globetrotting explorer and taxidermist. Akeley was a considerable influence in the museum world. First for the MPM, then for the Field Museum, and later at the American Museum of Natural History in New York, Akeley helped to elevate taxidermy from a craft that simply "stuffed" animal skins, to a blend of art and science. In addition to his technical contributions, Akeley, along with William T. Hornaday, argued that mounts of individual animal specimens were not enough; they should appear in groups as they did in the wild, in their reconstructed natural surroundings. Most

Figure 5. Twenty-one-year-old Herbert Stoddard with live gray fox while on a collecting trip along the Wisconsin River for the Milwaukee Public Museum. Photo courtesy of Leon and Julie Neel.

authorities consider Akeley's "Muskrat Group," completed for the MPM in 1889, to be the first fully realized habitat diorama.[35] Taxidermy, then, which began as a commercial trade to fulfill an urban public's desire to see nature, became a scientific pursuit that reflected an early shift from taxonomy to ecology.

Perhaps it is a mere coincidence that Akeley's break came via a deceased animal of P. T. Barnum, and Stoddard's one of the Ringling brothers, but it was no accident that Stoddard "early hitched [his] wagon to the Akeley star."[36] When Stoddard entered the field of museum taxidermy, Akeley was already a legend. He had not only revolutionized taxidermy and museum display—culminating with the African Hall at the American Museum of Natural History in New York—but he had also become a prototype for the great field collector and African explorer, that is, "the great white hunter." Tellingly, this image appealed to Stoddard. George Shrosbree worked with Akeley at both Ward's and the MPM, and he told Stoddard the stories that were so well chronicled in later years. These tales of Akeley's adventures "fired my imagination as have those of no other man, and I early decided to pattern my life after his as closely as possible."[37] Much like the Ringling Brothers' Circus had, Akeley's globetrotting and the taxidermic representations of what he encountered inspired in Stoddard a mystery and wonder about an unknown natural world. The African Hall, as historian Donna Haraway points out, "was meant to be a time machine" that transported the individual "to be received into a saved community." This was the intersection of taxidermy and conservation, a place where knowledge of the most threatened natural community might be preserved and disseminated.[38]

As fascinated as Stoddard was with the exotic nature of Akeley's displays, he was even more compelled to represent to the public a local nature that was quickly being altered by industrial America. Despite this alteration, though, there remained many "inner frontiers" close at hand in the upper Midwest, and they were critical to taxidermy display in the museums. It was through a studied realism that taxidermy would showcase disappearing nature. Stoddard put in long hours at the museum and hard study of natural history at night, and he quickly became known for his meticulous work in both the shop and the field. Like Akeley, he developed innovative taxidermy methods, especially in his bird work. He was likely the first to use electroplating—the coating of an object with a layer of metal—to best resemble a group of naked bird nestlings, and he was first to use cork and balsa wood for anatomical re-creation. And technique was crucial. A detail out of place would mislead the viewer and do an injustice to both science and nature.[39] It is important to note that Stoddard was one of the last to practice museum taxidermy before the rise of modern science devalued its status. Until the 1920s its practitioners still had designs on becoming a branch of the biological sciences. The Society of American Taxidermists (1881–83) was short lived and long dead by the time Stoddard entered the fray, but they had laid the professional groundwork for the field, which sought realism in both theory and practice.[40] As William T. Hornaday argued, "The task of the taxidermist, if properly appreciated, is a grave and serious one. It is not to depict the mere outline of an animal on paper or canvas. . . . It is to impart to a shapeless skin the exact size, the form, the attitude, the look of life."[41] The "look of life" for Stoddard reflected his growing interest in ornithology, and he sought to represent those scenes that were close at hand, yet hidden to most Americans. As much as he admired the work of older taxidermists such as Akeley—whom he met in 1918 in New York—Stoddard felt grand displays like those at the American Museum of Natural History obscured the smaller details of nature. During his visit with Akeley in 1918—while he awaited active duty in World War I—he became "a firmer believer than ever in the small detail groups after seeing the most elaborate and spectacular of the large groups in the country," and thought that other taxidermists "will all come around to it sooner or later."[42] The beauty and art of taxidermy, he thought, was its ability to expose what one may easily overlook in the nooks and crannies of a grand panorama.

On its surface, taxidermy as practiced in the early twentieth century was both a thoroughly urban and thoroughly artificial vision of nature. One seeking authenticity might simply see "stuffed" animals hanging on a wall or propped in a museum display case, completely detached from their natural context. This was mere representation and decontextualization, re-creating for urbanites what they were never likely to experience. But for museum taxidermists like Stoddard, the subject matter, and the process of creating the finished product, was more than detached nature. A completed habitat group represented a constellation of interactions with the natural world, a re-creation of what they knew intimately in the field. Many specimens came to the MPM and Field Museum through donation, but taxidermists themselves collected the vast majority, venturing afield not only to collect but also to study. Stoddard took meticulous field notes so as to make his habitat groups as authentic as possible. His specialty was bird groups, and when in the field he was careful to note bird behavior, relationships, and habitat.[43]

In his first few years in Milwaukee, Stoddard took several field trips throughout Wisconsin with Shrosbee, but he quickly lost patience with museum administration and left after only three years to take a job with the Field Museum in Chicago. The MPM, under the directorship of Henry L. Ward—the son of Henry A. Ward—had undertaken several steps toward modern efficiency that Stoddard felt had no place in museum work. Time cards, detailed reports of every minute spent on specimen preparation, and a disregard for accepted cataloging practices of scientific ornithology, made "the place seem like a factory." Ward, according to Stoddard, "had been influenced . . . by his background in Ward's Natural Science Establishment, from whence he came. This was a commercial institution, and operated as one."[44] Even museums, it seemed, could not avoid the efficiency studies of Frederick Taylor, who had published his highly influential *The Principles of Scientific Management* only two years prior. The Field Museum, on the other hand, maintained a loose work structure, allowing Stoddard to "make my own field studies and collect the material myself. . . . I carried on most of my field collecting in Illinois, Indiana, and Wisconsin; going, coming, and remaining in the field as long as I saw fit, and with no regimentation nor red tape."[45] Never one sympathetic to the human structures that limited his contact with nature, and rather more impressed with the designs found in nature, Stoddard flourished under such a system. Judging from his field notes, he spent as much time scouring the Lake Michigan shoreline as he did in the taxidermy shop. By the time he left the Field Museum to return to Milwaukee in 1920 (Ward had been relieved of duty), he had taken charge of all field expeditions for the bird groups. According to coworker Owen Gromme, "[Stoddard's] expedition work was strictly business with no time limits, and in his opinion anything that distracted a field collector beyond reasonable necessity was taboo."[46] His taxidermy work was first-rate, but it was in the field that Stoddard excelled. While a superficial viewing of "stuffed birds" in a museum hall might reflect a deadened or detached form of nature for many, to Stoddard it represented an intimate knowledge of the world of living birds as he found them in nature. His knowledge of how birds interacted in nature, and his experience of interacting with them in the field, made the act of building a museum display a representational process that reflected nature as he found it in the field.[47]

This kind of relationship with the environment in the early twentieth century was unusual. Stoddard was not a farmer or fur trapper, making a direct living off of the land. Though his work reflected the growing urban middle-class desire for interaction with nature, he was not a prototypical middle-class urbanite seeking respite and recuperation in nature. He was part of the burgeoning scientific class, a science worker, but not yet of the professional ilk attempting to set a natural resource agenda, or a university man devoted to research or constructing curricula. His employment with the museums provided an institutional home and scientific legitimacy, but museums, unlike government conservation departments, did not involve much bureaucratic infighting or constituency placating. In many ways, he co-opted the museums' institutional supports to pursue his own ornithological agenda. Most of Stoddard's fieldwork for both the MPM and the Field Museum involved collecting flora and fauna to be used in small bird groups. At the Field Museum, for instance, he worked in the Harris Public School Extension, where he built portable habitat groups that the museum loaned to the Chicago public school system. At a time when nature study and general

Figure 6. Herbert Stoddard gathering peregrine falcon nesting material in Sauk County, Wisconsin, for the Milwaukee Public Museum, 1921. As a taxidermist for the Milwaukee Public Museum and the Field Museum in Chicago, Stoddard spent as much time in the field observing and collecting as he did in the taxidermy studio. Photo courtesy of Leon and Julie Neel.

science were just securing a place in the public school curriculum, the extension's goal was to aid in scientific education through representation.[48] On the institutional level, this was not pure science; it was education. But while in the field Stoddard used his job as a taxidermist to rise in the ranks of the ornithological community, and his successes in both taxidermy and ornithology hinged on maintaining local connections in the field.

Throughout Stoddard's time with both the MPM and the Field Museum, Ed Ochsner continued to act as field contact. Two other Sauk County woodsmen—Alfred Gastrow and Bert Laws—also gave Stoddard invaluable assistance in the field. None of these men had much formal education between them, but to Stoddard they were among the most talented field men he ever knew. Through Stoddard's association with Laws, in particular, he "first began to draw a distinction between 'schooling' and 'education;' terms too often confused in American usage." Laws was a farmer who "actually cared little about farming," and was never known to read or write, but he "was the type who would investigate the construction of his crib as an infant, and display equal interest in the construction of his coffin at the end of the race. . . . I believe he learned something necessary or desirable every day of his life."[49] Writing in 1954, Stoddard was perhaps reflecting on his own life during a time when university-trained biologists had taken over his fields of expertise, but his reflections on Bert Laws help to explain his lifelong and stern emphasis on knowledge through experience in the field. Ostensibly writing on Laws, Stoddard thought "'schooling' *may* be an important part of 'education' in many or most cases. But I cannot swallow the general idea that a man cannot be 'educated' who has never attended 'school.' In some cases long-drawn-out school attendance may do more harm than good."[50] Absent the advanced schooling that Stoddard would never receive, his education came about instead through observation, notation, and trial and error application.

As important as field work was to Stoddard's education as an ornithologist and biologist, it also offered important lessons about nature study in a political world. State conservation commissioners, ornery landowners, and poaching fur trappers intermittently conflicted with goals in the field—particularly those aimed at collecting game animals. After one trip to the newly created Lake Wisconsin in 1922, the chairman of the Wisconsin Conservation Commission, W. E. Barber, dressed down Stoddard, Gromme, and Ochsner for misusing their scientific collecting permits. Barber notified Ochsner that the "sportsmen in the vicinity . . . are up in arms and we have received a gazing reprimand from that vicinity for the issuing of permits to be used as [they] were by you men." Locals witnessed the field party taking geese, canvasbacks, and other protected waterfowl out of season and were apparently unimpressed with their museum credentials. Barber threatened to "specify in each permit that no game bird is to be taken under said permit except in the open season for said bird."[51] Museum director S. A. Barrett took up the case, responding that it would be "entirely useless for us to endeavor to collect these birds during the open season, when all of the sportsmen are out and when the majority of the birds are in moult or in new plumage and quite unfit . . . for group building purposes or for scientific study."[52] After a meeting with the Conservation Commission and their state game wardens, Barrett was able to clear up the controversy, and Barber even "strongly requested that all game wardens throughout the State understand that the permits issued to the Museum's representatives were issued only after careful investigation and the positive assurance on the part of the Commission that such representatives were fully accredited, careful collectors who would never in any way abuse the privilege granted."[53]

Barber came around with some prodding, but the lesson that science was not practiced in a vacuum was clear. State commissions answered to a host of constituencies, many of which did not take kindly to an extralocal presence that seemed to operate outside of state and federal regulations. Locals themselves were just growing accustomed to the Migratory Bird Treaty Act of 1918, a treaty negotiated with Canada to set hunting regulations on game like ducks and geese.[54] Implementing such seasonal restrictions on waterfowl hunting, in particular, had been a hard-won victory for state and federal conservation organizations, and surely a bitter pill for many locals accustomed to hunting with few restrictions. If locals could not hunt game animals out of season, they would certainly raise their voices against those who could. Field scientists simply could not go about their work as if uninterested in the thoughts and actions of locals. Through incidents such as this, Stoddard gradually learned the importance of diplomacy in the field, apparently taking Barrett's advice to "take particular pains to avoid any action which can subject you . . . to the slightest criticism" regarding local affairs.[55] Such advice would be of primary importance a few years later when Stoddard arrived in the Red Hills.

Though Stoddard dreamed of circling the globe on collecting trips, the farthest he got—as a museum representative anyway—was Bonaventure Island, the famous gannet breeding ground in the Gulf of St. Lawrence off the coast of Quebec. Bonaventure became a protected bird sanctuary in 1919 as a part of the 1916 Migratory Bird Convention between the UK and United States to protect migratory birds in Canada and the United States, and the MPM sought to build a display "depicting the home life of the Gannets and other birds associated with them at their nesting colony there."[56] By the time Stoddard arrived at Bonaventure in July 1922, he had taken a conceptual leap in his interpretation and understanding of nature. Compared to the field notes of ten years earlier, those from the Bonaventure trip show a markedly more mature naturalist and ornithologist. The early notes are little more than listings of birds and mammals, a necessary first step in the world of taxonomic detail. In his later notes, the lists are still present, but they are embedded within passages of vivid detail about the process of collection, the mundane details of to and fro, an intuitive sense of the animals' relationships with their environment, and—most strikingly—a sense of awe and appreciation for natural beauty that is lacking in his earlier writings. As he approached the island with his local guide, Willie Duval, he saw

> a never-to-be forgotten sight. . . . The north and east sides come down to the water a sheer drop of 300 ft. and over, and in some places deeply undercut. Starting out to the northward and keeping as close as the sea allowed, we skirted the whole shore, first passing great cliffs and huge masses of fallen rock where the herring gulls nest by the thousands, and in half a mile or so coming to the gannet ledges. Ledge above ledge where every inch of available space is crowded with the magnificent birds,—row after row of soldier-like murres, with a few razor-billed Auks, puffins, Black Gullimot and Kittiwake gulls. . . . Late in the P.M. [Duval] and I circled the island, and enjoyed the amazing sight from above—a scene better recorded by camera than pencil.[57]

Indeed, the camera became Stoddard's favorite, if most unwieldy, tool to record bird life during these years, and despite his wish to use it to depict a panoramic view of Bonaventure, he more often used it to capture individual animal associations in the field. With both still and motion pictures, Stoddard could not only transport the dead birds and detritus of the field back to the shop, but he could also bring representations of how they were arranged in nature. As historian Gregg Mitman has shown, the camera became a powerful instrument for urban populations seeking more contact with nature.[58] Images from Bonaventure Island, and their replication by Stoddard's taxidermy work, would match perfectly with the aesthetic of wild nature sought by the MPM. Stoddard went about his duty on the monthlong trip, scouring beaches, rappelling off cliff ledges, noting bird behavior and habitat, taking still photos and motion pictures, all to gather material for the museum exhibit and place it in its natural context. But he also began to have a budding realization about nature's decline in the face of a rapidly changing world.

Perhaps Stoddard's sense of awe at Bonaventure is brought into greater relief in surroundings closer to home. The western and southern shores of Lake Michigan were among his favorite bird collecting and banding grounds, and by the early 1920s he feared for their inevitable transformation in the face of industrial and suburban growth. One section where the "prairies are entirely uncultivated and the original prairie flora still persists" was of particular concern. In 1923 he advised that "all data on the shore birds frequenting this strip of original prairie should be gathered next spring before it is too late. The growing industrial towns of Waukegan and Kenosha have already changed the character of much of this flat strip of lake shore. A recent real-estate development known as 'Chiwaukee' on the south border of that part of the prairie favored by the shore birds, points to further changes."[59] These areas that ecologist Frederic Clements called the "twilight zone between town and country" were quickly fading away, and Stoddard expressed little hope for the dunes and prairies of Lake Michigan writ large: "The whole western shore of Lake Michigan, from Green Bay on the north to the Indiana Dunes on the south seems to have been suddenly 'discovered.' Cottages and sub-divisions are springing up everywhere, and competing with the factories for the last remaining strip of shore line."[60]

Stoddard was not the only one concerned with the Lake Michigan shoreline. The Indiana Dunes was a place where ecological study came into its own, especially on the dynamics of vegetative succession, and by the early 1900s there were increasing calls for their preservation. A small group of scientists and reformers from Chicago organized advocacy groups and started leading tours to showcase the dunes' natural complexity. But in Stoddard's estimation, the dunes' renown as an ecological and scenic site was also its undoing. He was critical of the industrial and residential development that encroached on the dunes, but what seemed to bother him most was the increase in pedestrian traffic that followed, particularly those who viewed the dunes simply as a curiosity. Following a day in the field in 1919, he "was disgusted with my old favorite stamping grounds—all one could see for miles was . . . boys killing frogs and inadvertently breaking up every nest they ran across."[61]

Stoddard's contempt was not reserved for roaming boys. He also came down heavy on Henry Cowles, the University of Chicago botanist whose fieldwork in the dunes made him a pioneer in successional ecology. Cowles's 1899 paper, "Ecological Relations of Vegetation on the Sand Dunes of Lake Michigan," published in *Botanical Gazette*, brought international recognition for the dunes as an exemplar case study of early succession.[62] Cowles, like Stoddard, pushed hard in the early 1900s for preserving the dunes from commercial development, but Stoddard also believed such public awareness caught the attention of people who had little knowledge or appreciation of the area's ecological diversity. He wrote that in the 1910s, the Indiana Dunes "was a little-frequented and very diversified country, a paradise for the naturalist, though forgotten by everyone else. But that was before Dr. H. C. Cowles . . . wrote a glowing article for the National Geographic Magazine, telling of the wonders of the dune region. This was the beginning of the end as far as its wildness and splendid isolation were concerned."[63]

Stoddard probably meant to cite an article by another dune advocate, Orpheus Schantz, which declared the dunes "a national park opportunity," but he knew Cowles was one of the most visible proponents of making the dunes a park.[64] Given that they all wanted to see the dunes saved from development, Cowles seems like an odd choice on which to levy blame.

Stoddard's concern reveals a common line of thought among ornithologists about keeping the best birding ground a well-kept secret, but it also reveals a deeper apprehension about the nation's burgeoning conservation movement. Coupled with increased infrastructural and commercial development, preserving the dunes as a park would open the shoreline to the recreational use of Chicago's expanding consumer classes. It was fast becoming clear to him that a conservation intervention, when combined with the increased public exposure and the accompanying crowds of people, could be damaging to wild nature, or at least to the sensibilities of those like him who appreciated wild nature. As one of his correspondents would later write, reflecting Stoddard's own views, "While the Dunes Highway is a great convenience for motorists, I would gladly forego it in favor of the former seclusion of a region which, I fear, is more and more to become the despoiled and littered playground of thousands for whom its beauty and charm is without meaning."[65] Stoddard's experience with the Lake Michigan shoreline in general, and the Indiana Dunes in particular, demonstrates a scientific elitism that would continue to color his view of conservation. Wild places like the dunes were not only wild; they were laboratories in which to study the wild. Science workers like Stoddard discovered and nurtured these representations of wildness, but they had little interest in putting them on display for casual observers.

Stoddard's experience as a museum field man and taxidermist led to a deepening involvement in the ornithological community, where he made connections critical to his later return to the Southeast. He joined the American Ornithologists' Union in 1912, and within a few years was member of the Wilson Club, Cooper Club, and National Audubon Society, all national organizations that welcomed both amateur and professional ornithologists. Within a short time after returning to Milwaukee in 1920, his reputation in ornithological circles was such that his coworker Owen Gromme thought "he was unquestionably [Wisconsin's] foremost ornithologist."[66]

Stoddard first heard of a possible southern quail investigation at the 1922 meeting of the AOU, where he helped to found the Inland Bird Banding Association along with legendary bird-bander S. Prentiss Baldwin. Bird banding had been used as a research technique since the turn of the twentieth century to assemble data on migratory patterns. But not until Baldwin's innovative trapping methods—which used government sparrow traps to band and track adult birds instead of banding nestlings in the hope that the band would be returned upon death—did it become a truly effective research tool. Between 1914 and 1920 Baldwin—a Cleveland businessman and one of the most influential amateur ornithologists of the era—banded thousands of birds at his summer home outside of Cleveland and at his winter getaway in, of all places, Thomasville, Georgia. With the advent of banding, ornithology historian Mark Barrow explains, "individual birds could now be recaptured multiple times at various locations, thereby tracing their movements over time."[67] With a feasible way of trapping adult birds, the U.S. Bureau of Biological Survey took over sponsorship of the program and promoted and trained individuals across the nation to band birds.

Stoddard was among the first to follow Baldwin's lead into banding. In 1923 he wrote in the *Yearbook of the Public Museum of the City of Milwaukee* that bird banding "bids fair to revolutionize bird study."[68] He banded birds throughout his travels for the MPM and began publishing accounts of his field experience in journals such as *The Auk* and *The Wilson Bulletin*, drawing the attention of a wider audience.[69] Baldwin, along with W. L. McAtee of the Biological Survey, soon took note of Stoddard's field work for the museum and the Inland Bird Banding Association, and kept him in mind for any possible job openings in the survey.

Stoddard came to ornithology as it became a trend-setting field in conservation and science. As has already been noted, it was accessible, resisting academic professionalization far longer than other fields in the biological sciences. Even as graduate programs surfaced throughout the early twentieth century, it retained strong popular appeal. In large part, such popular interest came in reaction to the earlier exploits of market hunters, who ranged far and wide with little concern for the depletion, and in some cases extinction, of bird species. One result of this assault was government action. Congress overwhelmingly passed the Lacey Act in 1900, which prohibited the transportation of illegally taken wildlife across state lines. The Biological Survey set up an informal network of wardens to help enforce the Lacey Act, and took on greater responsibility for tracking migratory patterns and studying breeding and food habits. The Biological Survey, originally formed at the instigation of the American Ornithologists' Union as the Division of Economic Ornithology and Mammalogy in 1886, was a bureau of the federal government's burgeoning agricultural research establishment, and it became crucial to the nascent conservation movement of the Progressive era. From the beginning, the survey had to balance the wishes of Congress that it aid farmers, and the desire of many of its employees to conduct biological research. As it grew throughout the early 1920s to become a broad-based government division focused on general wildlife study, the survey was part of a coarsely woven network of academic departments, natural history museums and societies, and the expressly practical sciences of the U.S. Department of Agriculture (USDA) and Forest Service. By the 1920s its administrative and research capacities were stretched thinly across a broad range of wildlife-related issues. Its core constituency was ornithologists, but it was increasingly called upon for administrative and research support by mammalogists, state game agencies and law enforcement, and recreating sportsmen.[70]

When approached by a group of Red Hills sportsmen, Bureau chief E. W. Nelson gladly granted them a hearing. On April 25, 1923, a small group of landowners from the Red Hills met with a representative of the Bureau of Biological Survey at the exclusive Links Club in New York City. There, they discussed the possibility of an investigation into the life history of the bobwhite quail in the Southeast. It was not unusual for sportsmen to call on the Biological Survey for help with game problems, but this particular request was curious. The preoccupations of most sporting groups up to this point typically revolved around hunting seasons, predator control, or wildlife ownership. But here was a group of sportsmen calling for a life history study. The life history concept arose from ornithologists frustrated with simply collecting specimens or tracking movements, rather than actually recording the habits of birds in the wild. Several researchers in the egg-collecting branch of ornithology (oology) attempted life history studies around the turn of the century, but not until the 1910s did they gain wider acceptance.[71] The Red Hills sportsmen were most likely acting on the advice of Prentiss Baldwin. With a foot in the worlds of both sportsmen and ornithologists, he recognized the two interests could converge to push the life history concept beyond its limited role as a descriptive addition to ornithological taxonomy. Instead, the life history could be a prescriptive conservation tool designed to produce applicable results. The data compiled would not simply sit on a shelf to be referenced by professional ornithologists; it would be applied to the development of practical solutions for land managers in the field. The major thrust of the proposed quail study was to explain the recent decline in quail, to discover what environmental conditions best suited the birds, and then to make management recommendations based on such discoveries. Ultimately, the preserve owners and the Biological Survey would create the framework for the most thorough and innovative life history of a game species up to that point.

Within a year of the Links Club meeting an agreement was signed between the preserve owners and the Biological Survey. Baldwin and McAtee quickly agreed that Stoddard had "the qualities we think are desirable for the position combined in one man."[72] Those qualities apparently did not include a high school diploma or any real experience conducting what promised to be a thorough scientific study. One problem for the survey was a dearth of qualified leaders for the investigation. There were a handful of candidates at various state game farms, and a few university researchers already engaged in less practical animal research, but very few who had Stoddard's field experience.[73] More than a trained and educated scientist, the Cooperative Quail Investigation needed a self-starter who would be able to improvise in the field and work efficiently under a mere skeleton of centralized directives, qualities that S. A. Barrett, director of the MPM, had already instilled in Stoddard: "The conditions which you encounter in the field must govern your actions and your own judgment is about the only thing that can count for much when it comes to field work . . . when we send a man out into the field we depend upon him to use his judgment and to secure the best possible results and do not wish to dictate from the office just what you shall do."[74] It was just such ability to handle the intangibles that ultimately recommended Stoddard to the Biological Survey. Like so many early federal projects in the field, no one really knew what this one would look like; as Biological Survey chief E. W. Nelson put it to Stoddard in the job offer, "the success or failure of the investigation will rest in your hands, and the initiative will rest largely with you as to how the work is carried on and the results obtained."[75] Again, this was to be a new type of study regarding wildlife. Forestry and the agricultural sciences had long been engaged in conducting practical research to aid in production, but there had been no real effort to regulate natural processes in the hope of aiding wildlife. There was simply no methodological formula to follow.

Stoddard immediately accepted the challenge. His boss at the MPM, S. A. Barrett, conceded that "the offer made Mr. Stoddard by the Biological Survey is so much in advance of anything that we could offer him here. . . . [T]here is nothing left for him in justice to himself and his future but to accept this proposal."[76] When he joined the sponsors of the quail study in Thomasville in February 1924, he had left in the Upper Midwest many friends and professional colleagues whom he remained close to for many years. Ironically, he left the Upper Midwest just as another son of that region, Aldo Leopold, returned to it. Leopold would, over the next couple of decades, become intimate with some of the very ground where Stoddard had honed his skills as a naturalist. As will become clear in the following chapters, he and Stoddard would also greatly influence each other and become close friends and colleagues.

Stoddard came to the Biological Survey through a series of fortuitous circumstances and events. His time in the old-growth longleaf woodlands of central Florida with cattle ranchers fostered a deep interest in natural history; his family's financial status back in the Midwest required that he make a living, thus leading to an apprenticeship with Ed Ochsner; Ochsner's friendship, in turn, gave Stoddard entrée into the profession of taxidermy and a web of personal relationships cultivated in the museum world; and finally, these relationships led to notoriety in the ornithology community and the appointment as head of the Biological Survey's quail investigation. Not only were the circumstances of his professional growth important, but the places in which he learned were as well. By 1920 these places contained both wild and working landscapes within close proximity, as well as the threat of industrial, commercial, and infrastructural growth. Within this matrix of disturbance, science workers like Stoddard found places to learn about the natural world. As he moved on to study the life history of the bobwhite quail, he was following an important trend in the biological sciences away from descriptive taxonomy toward explanatory ecology. But Stoddard's practical background in field biology required that he interact with nature; he would never approach the study of ecology in abstract theoretical terms. Instead, his explanatory device would become the land itself.

CHAPTER 3

Putting Fire in Its Place

As he did most every day while visiting the Red Hills region of south Georgia and north Florida, Henry Beadel—the son of a northern industrialist—was quail hunting with his brother, Gerald, and their African American driver, Charley. It was a chilly afternoon in February, late 1890s. Upon reaching their shooting grounds, Beadel witnessed the unthinkable: "We saw the whole country on fire, which within a few minutes left the ground black and bare except for scattered clumps of bushes." An area that only the day before stood as an idyllic scene of grand pine woodlands interspersed with small, almost meadowlike agricultural fields, now appeared before them as a fire-blackened hell-on-earth. Unbeknownst to Beadel, the local African American sharecroppers had "put the fire out" that afternoon, ridding field and forest of a year's worth of accumulated growth. Beadel was not amused. "The country looked to us irretrievably ruined, and the quail *doomed*."[1]

Charley soon set Beadel's mind at ease. He "informed us that this burning took place regularly every spring as far back as his great-grandpapa could remember." Relieved, yet still a bit incredulous, Beadel took "a few calmer squints through the smoke [to see] all the trees still standing, and we even found that we could walk behind the flames without scorching our boots." After a little sleuthing, he discovered that locals "took the practice as much for granted that it had not occurred to them to mention it to us."[2] Setting fires was one of the many local land management practices that mimicked historical ecological disturbance in the South's longleaf-grassland environment—practices that would soon be repeatedly attacked and defended by a bevy of scientific experts.

Almost three decades later, it did not take quite such a revelatory experience for Herbert Stoddard to realize fire had an essential place in the South's coastal plain ecology. Despite the anti-fire dogma that forestry experts spread around the region in the 1920s, he had a strong inclination before arriving in the Red Hills that the stability of the region's longleaf pine woodlands depended on routine fire. Stoddard arrived in Thomasville on February 7, 1924, and immediately began his immersion in the Red Hills environment. His exploration started on the sprawling Thomas County properties of Charles Chapin, Howard Hanna, and Jeptha Wade. Riding with Chapin on February 9, he recorded in his field diary that he saw a "world of beautiful quail country. As a rough guess I should say that about 75% of the country is covered with timber (and much of this so scattered and open that it is *ideal* quail country)." There were many patches of corn planted throughout the forests and a "considerable acreage [of] cotton land," but it was the woodlands that got his attention. He marveled that "the majority of the pine land is primeval woods which has never been cut over or turpentined"; and like so many travelers before him, Stoddard noted that "this section is all rolling and very beautiful country, in marked contrast to the flat, desolate turpentined to death country to the east." The health-seeking landowners had gotten to these lands before the timber industry could do its work, and the forested grasslands remained largely intact. He seemed unfazed that "the majority is burned over on an average of every two years." He remembered well the annual ritual of burning off the cattle range in central Florida, and like his childhood home, here in Thomas County "most of the wooded country is almost a perfect stand of yellow or 'long leaf' pine." The presence of fire, then, seemed to make sense, or at least it did not raise any immediate concerns.[3]

Many people had likely taken a similar route as Stoddard did that day, but few recorded what they saw in such detail. Unlike so many previous travelers to the Red Hills who were escaping work, Stoddard was in the midst of it, conducting his initial survey of the properties as an agent of the federal government's Bureau of Biological Survey. Before Stoddard's arrival, the owners and residents of the quail preserves had gone about their business largely outside the province of the government. Hunting seasons and bag limits were in place, but beyond these regulations there was little government direction or interest in how the preserves carried on their affairs. Stoddard's arrival signified something new. He was there in a governmental, and scientific, capacity to help the preserve owners create a system of land management to enhance their quail shooting, but it is striking how little the land-use patterns actually changed under Stoddard's direction. Instead, Stoddard adapted many practices of southern coastal plain agriculture to help create a hybrid form of land management based in both the biological sciences and the practical experience of production. Rather than forcing a cultural landscape to fit an abstract set of scientific principles, he set out to mold a system of management from the region's cultural and environmental past. In addition to being a land of both respite and work, the circumscribed environment of the quail preserves was now an open-ended scientific laboratory.

The remaining chapters examine the process through which Stoddard gained scientific knowledge about the longleaf pine woodlands of the Red Hills, and how that knowledge was, in turn, applied and disseminated as a conservation alternative in the longleaf pine–grassland region. This was not theoretical field science for its own sake; the goal was always practical application. As Robert Kohler has argued, field biologists "do not just work *in* a place, as laboratory biologists do, but *on* it. Places are as much the object of their work as the creatures that live in them."[4] Stoddard came to study the life history of the bobwhite quail, and to do so required study of its place, its dynamic environmental surroundings. Furthermore, Stoddard came to realize that the Red Hills environment existed in delicate tension with the complex cultural world of humans. The accumulation and application of scientific knowledge in the field, then, was not a simple endeavor. Stoddard faced a host of natural and cultural disturbances in the field, and quite a few combinations of the two. Hurricanes, drought, cotton agriculture, domestic pets, and exotic plant and animal pests were only a few of the difficulties he dealt with locally in pursuit of quail production. In other words, there were a host of uncontrollable variables that affected scientific fieldwork. Dealing with so many variables required a predilection toward improvisation, as well as a measure of common sense.[5]

For the science of game management, the Cooperative Quail Investigation (CQI) was a landmark study. It helped to define and propagate the field of game management—which was also intermittently called wild life or wildlife management throughout the 1930s until the formation of The Wildlife Society in 1936 secured the place of nongame species in the profession—and helped make the field one of the most important among a growing number of natural resource professions.[6] The book that resulted from Stoddard's investigations, *The Bobwhite Quail: Its Habits, Preservation and Increase* (1931), is often overlooked by environmental historians today, but conservation advocates of the day regarded it as the premier document on wildlife management of any species in any region. And while Aldo Leopold is often considered the founder of wildlife management, Leopold himself considered that honor to be Stoddard's. The two friends worked closely together to outline the field of wildlife management, both in the field and through correspondence, but it was Stoddard who designed and implemented the first model study. Leopold's biographer, Curt Meine, notes that Stoddard was "the first to examine a game species in detail and to utilize that information in a restoration effort. While Leopold was evolving an abstract framework for the science, Stoddard was providing its first concrete example."[7] Leopold himself was even more generous, giving Stoddard credit for blending the abstract with the concrete. While Stoddard put the finishing touches on *The Bobwhite Quail*'s galleys, Leopold predicted it would

> set an entirely new standard which will appeal, I think, to both the fundamental biologist and the agricultural administrator when they become acquainted with it. Stoddard's Georgia Quail Investigation has set the pattern for the new method. His report, just coming off the press, will illustrate my meaning better than I can explain it. His technique consists of an alternation of field observation to get "leads," and controlled experiments to test their validity. As far as I know, his was the first attempt to weigh all the factors which determine the abundance of an American game species. He ends with almost an equation for the abundance of quail. This, as nearly as I know, is fundamental ecology. On the other hand, he also ends up with a distinct system of practice for the landowner. This, as nearly as I know, is applied ecology or agriculture.[8]

Despite Stoddard's narrow title, then, this was not simply a study with a singular focus on one animal species. In his treatment of the bobwhite quail, he examined patterns, connections, and associations within all of nature, and argued that in the face of rapid industrial transformation, the intentional management of landed resources for the purposes of ecological diversity was not just feasible but essential to the survival of a variety of plant and animal life. *The Bobwhite Quail* was thus a seminal text, and Stoddard a pioneering figure, in the early study of wildlife ecology and management.

The Bobwhite Quail addressed countless details of quail life history and management, but its findings on three subjects in particular would deeply influence land management policy on public and private land in the longleaf pine region and beyond. First, Stoddard concluded that fire had a natural role in longleaf pine woodlands, and that the intentional application of fire would maintain natural conditions best suited for quail and other wildlife, as well as perpetuate the stability of the forests. Second, he found that most quail predators had a negligible effect on quail populations, and that environmental control of predators was more effective than attempts at eradication. Third, he argued that the growth of intensive, mechanized agriculture constituted the greatest threat to wildlife resources across the South and the nation. These three findings ran counter to many commonly held assumptions of scientists and practitioners in several fields of study, and Stoddard arrived at them just as a handful of others across the South and nation began to ask similar questions. The acceptance of fire was the first finding to draw national attention to the CQI, and it propelled Stoddard into the unfamiliar role of public policy critic.

After Herbert Stoddard's weeklong survey of the Red Hills in early February 1924, he returned in March to begin the quail work in earnest. He and his wife, Ada, and son, Sonny, set up residence at Lewis Thompson's thousand-acre satellite plantation in Grady County, Georgia. Known as "The Hall," and later called Sherwood Plantation, this piece of land served as the investigation's research base, along with Thompson's fifteen-thousand-acre Sunny Hill Plantation. Sherwood's provenance was typical of the quail properties. It was an antebellum cotton plantation, and then, after the Civil War and Reconstruction, it passed to J. J. Healy, a priest from New England who purchased it as a winter retreat. When Healy's failing health no longer allowed him to travel south, H. W. Hopkins bought Sherwood in 1906 and used it for hunting and raising Shetland ponies, sheep, and hogs. Hopkins, though, rarely held a piece of land for long, and Sherwood was no different. Lewis Thompson purchased it a few years later and used it as a sort of retreat within a retreat. When he tired of hunting the countryside of Sunny Hill, he packed up and spent a few days at the more heavily wooded Sherwood. When Stoddard moved in he described its landscape as "largely covered with open pine woods. About 60 to 100 acres has been cleared in three or four small patches. This has been planted mainly to corn, some cotton. There is a beautiful clear 'branch' running across the place, bordered by many very large pines, some beautiful magnolias and deciduous trees."[9] Though smaller in size than most hunting preserves, both Sherwood's lineage and its environmental components were consistent with the surrounding preserves.

The CQI was a cooperative project between the preserve owners and Biological Survey. The preserve owners organized a group of about seventy-five contributors, mostly from the Red Hills, who funded the entire study and offered their land base for observation and experimentation. The survey provided Stoddard with institutional and laboratory support, especially in the identification of avian diseases. They sent Charles O. Handley, a field agent in the food habits division, to aid Stoddard in the field, and the two of them immediately set out to examine the Red Hills environment, trapping and banding quail; observing breeding behavior, nest sites, and daily behavior; examining the anatomy of quail and their predators; and observing predator-prey relations.[10] By this time, the quail preserves encompassed about three hundred thousand acres—much of it contiguous—and Stoddard had access to practically all of it.

In these early stages of the study, Stoddard was exceedingly careful not to offend the considerable egos of the preserve owners. He did well in this regard, but not without some coaching from S. Prentiss Baldwin, who spent many winters in Thomasville as an ornithologist and had some experience with the sportsmen. First, Stoddard had to establish his identity as a scientist. His status as a field agent of the Biological Survey provided him a certain amount of scientific legitimacy, but both Baldwin and E. W. Nelson were concerned about how the preserve owners might interpret his lack of formal training. By this time, taxidermy was losing favor as a reputable field within the sciences, and Baldwin thought the preserve owners needed reassurance "that you come to them as a scientist." To do so, they advised Stoddard not to overtly collect bird skins or practice taxidermy, as "it would not do to let the quail men think of you as just a taxidermist."[11] Baldwin himself had long been trying to convince his neighbors in the Red Hills to "appreciate that I am known in science," but he admitted that "the all important thing is not what they think of me, anyway, but that Thompson and the others have absolute confidence in you and the Survey."[12]

Indeed, Lewis Thompson was the most active preserve owner among the investigation contributors, and his cooperation and confidence would be crucial. The son of William Payne Thompson, a West Virginia oil producer who became a director of Standard Oil, Thompson made his home in Red Bank, New Jersey, but spent his winters at Sunny Hill and the rest of the year traveling from the Florida Keys to the Pacific Northwest chasing game and fish of various stripe. Unofficially known as "the finest shot of any man in the country," he participated in shooting matches across the country, was active in New Jersey's Republican Party, and one year followed the duck migration from Nova Scotia to Mexico, extending his own personal duck season to four months.[13] He was accustomed to getting what he wanted—according to Baldwin, he "at times takes more drink than is wise and is sometimes more forcible than correct in his thinking."[14] Despite his brusque manner, or his penchant for control, Thompson immediately took a liking to Stoddard. He appreciated, according to Baldwin, that Stoddard was "modest, and sincere, and willing to learn (not claiming to know it all). . . . You should have heard Thompson telling me just how [the study] ought to be done; and all of it just what I knew he got from you."[15]

Thompson's presence kept Stoddard on his toes, but alongside the local land managers and tenants with whom he worked every day, he could be a bit more casual. Unlike many agents of the nascent conservation state, Stoddard was not seen as an outsider by locals. Stoddard's rearing more closely approximated that of the local preserve managers and laborers, and his experience with people like Ed Ochsner and Bert Laws in Wisconsin gave him an appreciation for those who spent their lives in a particular environment. Stoddard's most trusted companions and informants during the study were local preserve managers. Among those who befriended him early on were Louis Campbell, then the dog handler and later manager of Harry Payne Whitney's Forshala Plantation; Robert Stringer, manager of Thompson's Sunny Hill; and Stringer's sons, Sidney, Albert, and Robert Jr. Thompson's employees were particularly helpful to Stoddard. Though he ceded all authority to Stoddard for running the study, from day one Thompson instructed his managers and tenants "to aid and assist [the study] in every possible way and manner, and later told me to never hesitate to ask for anything I might want or need to further the work."[16] This largess provided important support for Stoddard's work, but the giving of both human and land resources also meant Lewis Thompson retained considerable control over the activity and progress of the quail investigation.

This reliance on locals made it impossible, even undesirable, to conduct the quail investigation in isolation from the community. Stoddard engaged most everyone he met in town and country with discussions on quail and was quick to capitalize on the help offered by Thompson and others. Much of his received advice came from the preserve owners, but Stoddard was more interested in what natives had to say because "local people were familiar with the birds in summer," rather than only during the hunting season.[17] Stoddard wrote that "the whole terrain is daily crisscrossed by the trails of people going about their work afield," and it was not long before the area's small farmers, managers, tenants, and wage hands fielded requests of all types from Stoddard.[18] As the ones who worked the fields and forests on a daily basis, locals provided important information on what they considered normative and anomalous environmental conditions.

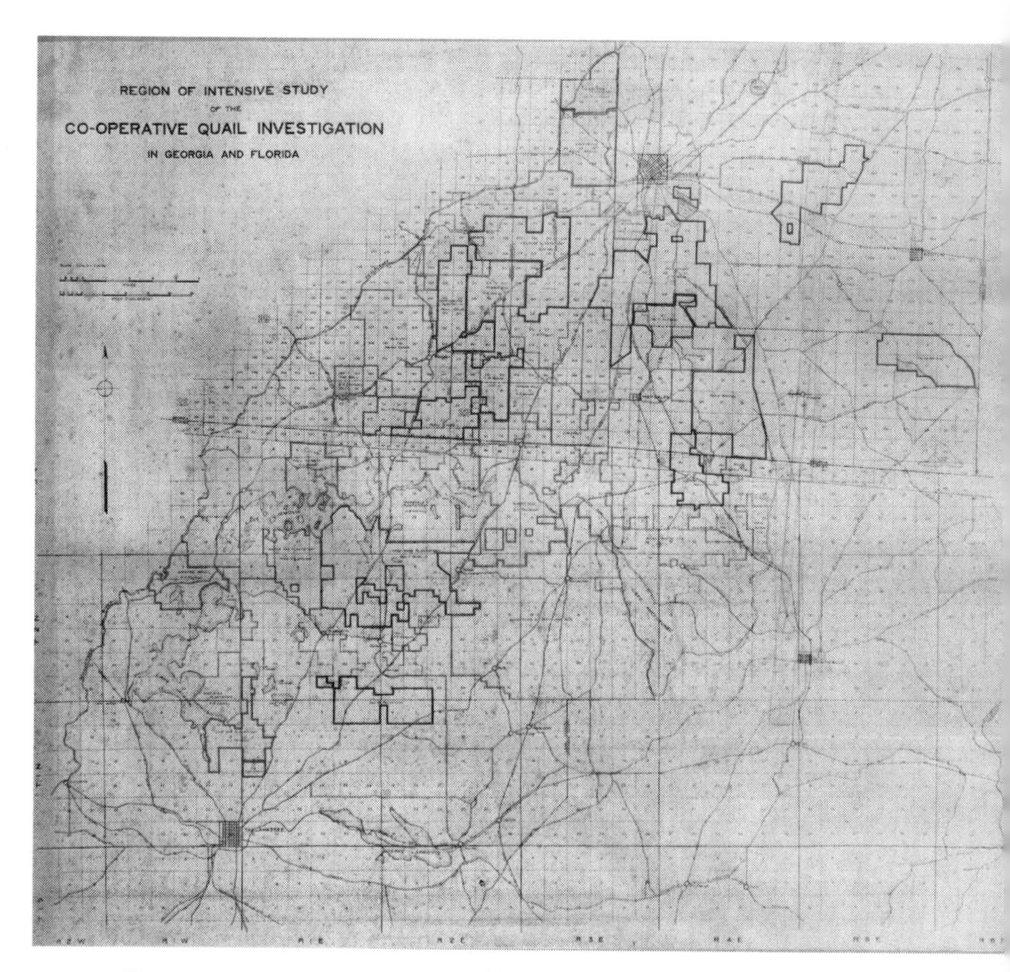

Figure 7. CQI Map, 1924. Among Stoddard's first duties with the Cooperative Quail Investigation was to create a detailed map of the Red Hills and its hunting preserves. Note the outlines of each preserve and the grids of Thomasville and Tallahassee in the upper right and lower left corners, respectively. Courtesy of Tall Timbers Research Station.

When it came to the doings of quail, and in particular the reasons for their decline, Stoddard soon learned that everyone had an opinion. He lauded eighty-five-year-old H. W. Hopkins as "an inexhaustible fount of local information, and an immense amount of practical information about the quail." Over the course of several early visits they "talked over quail, quail-shooting, [and] trapping methods," and Hopkins confided that "he has studied them for over fifty years and has come to the conclusion that he really *knows* very little about them!—which proves he is a very wise man."[19] Other opinions varied widely about what caused the decline in quail, and Stoddard noted most of them in his field diaries. One Leon County native thought an increase in rattlesnakes spelled quail doom; another considered overshooting to be the culprit. During one trip south of Tallahassee while taking shelter from rain, Stoddard was sitting with locals when "the conversation turned to quail and the dozen or more all gave their news. House cats were considered as the worst enemies by all present. . . . One motherly old lady said that her own cat had already 'cleaned out two coveys of young quail' that frequented the immediate vicinity of her home!"[20] This surely rang true with Stoddard—house cats are not usually a favorite of ornithologists.

Many others, however, did not regard quail numbers to be on the decline at all. One contact thought "they find almost as many coveys now" as in years past, but because of increased hunting pressure "quail are gradually changing their habits and becoming harder to shoot, being found less in the broom fields and open ground than formerly."[21] Stoddard himself thought the "scarcity to be more apparent than real," but he also knew that a host of uncontrollable climatic factors had a hand in determining seasonal fluctuations.[22] Stoddard took these sometimes conflicting opinions seriously, and used many of them as the basis for his experimentation and observation. He also believed in the efficacy of the scientific method, and through it sought to confirm the truths and debunk the myths. In this way, the quail study became a very public activity. Everyone, resident and nonresident alike, had an interest in quail, and though some locals "appeared a trifle skeptical as to what we would learn," most welcomed the study's aims and considered it a public good.[23]

The more Stoddard saw and heard, the more he realized the need for a guiding hypothesis to drive his research agenda. What he devised was simple—as well as flawed in its earliest incarnation—but it led to basic questions in wildlife management that directed the field for decades to come, and eventually turned some basic assumptions of forestry and agriculture on their heads. His most important step toward formulating a hypothesis was to connect quail to their surrounding vegetation. He proposed that if targeted environments contained enough food and cover, that is, habitat, then quail would "breed up" to reach a maximum capacity. A host of complications might arise to prevent reaching the range's capacity, but this maxim generated the core questions of the study: 1) what were the preferred foods of bobwhite quail? 2) how can we make more of them grow? 3) what were the most prevalent quail predators? 4) how can we see that predators do not become a nuisance before quail populations reach maximum capacity?

In essence, Stoddard wanted to determine the "carrying capacity" of the quail range and develop methods to enhance it on an annual basis. The concept of carrying capacity had been used to navigate range management issues in the West and had only recently cropped up in Western big-game management issues, but any practical measures to strike a balance between food supply and animal populations in the West would have little bearing on management in the longleaf woodlands.[24] When Biological Survey chief E. W. Nelson first visited the Red Hills in April 1924, Stoddard presented some initial thoughts on "the stimulating effect of an abundant food supply on animal life in general." He was pleased Nelson agreed that "there is no doubt that the tendency of animals was to attempt to 'breed up to the food supply' and that an abundant and nourishing food supply undoubtedly would serve to increase quail as it did other creatures."[25] The difficulty, and promise, of such an assumption lay in rooting out the mind-numbing ecological complexity of how animal populations related to one another, as well as to the vegetative associations of the longleaf woodlands.

The first problem to work out was that of habitat—the mix of food and cover vital to quail survival. In determining what food plants quail preferred, Stoddard observed feeding habits in the field, examined countless quail crops and gizzards, and built a quail propagating plant on Sherwood to manipulate the feeding habits of captive birds. He also experimented endlessly with various quail food plants to determine the best seed collecting, planting, and cultivation techniques. The results enabled him to advise landowners on what to plant to increase their quail supply. More importantly, he concluded that many traditional land management techniques were already in place to ensure abundant food and cover. In the expansive longleaf woodlands, the use of fire ensured the annual renewal of wild seed-bearing plants crucial to quail, and the small-scale patch farming of tenantry created edge effects that provided both food and cover.

Figure 8. Herbert Stoddard banding bobwhite quail, 1924. Stoddard was one of the early proponents of bird banding as a research technique to track bird movements. Baited traps such as this one were most effective for capturing wild bobwhite quail during the CQI. Note the quail in Stoddard's left hand, the pliers used for clamping the band on the bird's leg in his right hand, and the ever-present field diary opened on top of the trap. Photo courtesy of Leon and Julie Neel.

Within the first two years of the study, Stoddard's observations of plant growth in the field led him to what would become a revolutionary conclusion in southern resource management circles: the best way to increase the food supply, and thus increase quail populations, in the longleaf pine woodlands was to regularly burn the forests. In such a fire-adapted system, grasses and legumes like Florida beggarweed, partridge pea, and bluestems would quickly re-sprout after a burn, thus providing quail and other wildlife a constant food supply. Without fire, these vegetative associations would be quickly shaded out by hardwood growth. Years later, when his land management system was thoroughly developed, he advised one landowner who wondered how to plant more partridge pea that "for the price of a box of matches all of your properties in both Florida and Georgia are kept full of this fine feed by the controlled burning carried on by your men."[26] Why rely solely on planting, he thought, when you could simply burn the woods and create ideal conditions for natural feed plants to grow? In the longleaf pine woodlands, then, fire formed the cornerstone of Stoddard's practice of science and land management.

It did not take long for Stoddard to realize that the study of fire's effects on flora and fauna would be a major part of the investigation. He had little choice: it was a cultural and material reality that residents of the quail preserves burned most of the farm and forest land every year or so in late winter and early spring. In some ways, the burning habits that Stoddard encountered were simply part of a southern mimetic tradition, passed down by word and deed through the generations. The material reasons for using fire, however, were part of the physical reality of living in the longleaf forests. As fire historian Stephen Pyne has pointed out, "The fire history of the South is in good part a history of its fuels."[27] In the longleaf woodlands, fuels constitute the mass of understory grasses and woody debris—often called the "rough" in southern parlance—that accumulate particularly fast when not checked by fire. After only a few years without a fire, such unchecked growth made living in the southern piney woods difficult for both humans and many species of wildlife, and greatly increased the chances of a large conflagration during periods of drought.

The earliest inhabitants of the region were quick to pick up on the beneficence of fire. Early settlers in the South borrowed the burning habit from Native Americans, who used fire to clear agricultural land, lure game animals, and keep forests open for their own comfort and security. As settlers spread across the region in the eighteenth and nineteenth centuries, burning became the dominant land management tradition throughout the region. Farmers used it to open new land and to maintain older fields by removing postharvest debris. Cattle herders burned the open range yearly to encourage nutritious new growth in the unimproved coastal plain uplands. Naval stores operators burned their woods to facilitate the movement of labor from tree to tree in gathering turpentine, as well as to control ticks and chiggers, and to make rattlesnakes more visible.[28] Until the early twentieth century, virtually every land use in the piney woods South required some sort of fire application. The movement of industrial forestry into the South around the turn of the twentieth century called many of these motivations into question, but when Stoddard came to the Red Hills there was considerable persistence in setting the fields and forests aflame.

Stoddard's childhood experience with fire, his discussions with local land managers like Campbell and the Stringers, and his observations of tenant burning methods convinced him that fire was a useful tool to make the forests more manageable for human use. His earliest observations noted that February "seems to be the height of the burning season and considerable stretches have been burned over," and that it was the local policy to "burn every few years, otherwise the woods became impassable to wagons and difficult to hunt."[29] Beyond making the forests more congenial to social and economic activity, however, fire's ecological role was initially less clear to Stoddard. Early in the study he took care not to assume that fire was a necessary natural and cultural component in the longleaf forest. Upon first glance, he thought "the quail must move out of newly burned country," and he was concerned about the effects of one fire he observed that "was very hot in spots for it killed the leaves of scrub oaks for 40 ft up, and would have destroyed any dried [songbird] nests in these trees."[30] These early concerns were likely those of a cautious young scientist, for the longer he observed burned-over areas, the more certain he became that fire, as it was used in the Red Hills, had little detrimental effect on the forests. One afternoon he found a covey "dusting in an open situation in a huge burn, fully 400 yards from nearest cover. . . . This is further evidence that many coveys continue to live on the burn, or in tiny patches of cover that many be left here or there."[31] A few days later when "fires [were] raging everywhere. . . . The burned areas seem to be the place to look for quail, as burned pine woods seem to be greatly favored for feeding and little patches of vegetation that escaped the burning are used for sunning, dusting, and to some extent, for roosting."[32] Quail seemed to relish the ash left behind, but what became of the plant life that fueled the fires and contained the core nutritional content for quail? The bare mineral soil left in a fire's wake would not sustain a quail population for long. It was here, in his study of understory plant life, that Stoddard began to develop a more complete understanding of fire's natural role in the longleaf-grassland forest.

Though the majority of preserves burned the entirety of their forests every two years, a few curtailed the practice several years before Stoddard began the study, thus making for ideal comparative observation of ground cover. Jeptha Wade, for instance, on the advice of landscape architect Warren Manning, had only burned sporadically on his Mill Pond Plantation since around 1906. Wade found the results less than satisfactory. He told Stoddard that "they have burned very little for years, but the place was getting so brush choked that they burned it off last year for the first time." Throughout his policy of fire exclusion, Wade noticed "a marked falling off in their quail supply."[33] The rough became a dense mix of hardwood reproduction, vines, and tangles often described as "jungle-like," which made the forest not only difficult for humans to traverse but also choked out important food plants for quail and most any other wildlife commonly found in the uplands of the southern coastal plain.

Comparing Wade's unburned rough with the lush forest understory found on other preserves gave Stoddard some obvious contrasts to work with. On Lewis Thompson's Sunny Hill, for example, tenants continued a heavy burning rotation. Stoddard inspected one section known as the "Perkins woods" in May 1924 that "had been burned over early in the season, and the new vegetation had acquired a good growth" within just a few months. The Perkins woods was "so open that there is a dense ground growth of a wide variety of plants" and residents there "tell me that it is shot hard 7–8 times a season, but that they always find birds."[34] Over the first two years of the study, as he watched a diverse assortment of plant life return on burned land and disappear on unburned land, Stoddard came to understand that the parklike woodland landscape he first encountered in the Red Hills—that aesthetic so attractive to quail hunters—was actually the result of frequent fire. This likely came as no great surprise to locals in the region, nor did it to Stoddard, but local common sense did not easily translate into verifiable science.

Nevertheless, when it came time to draw up the first substantive report on the quail investigation, Stoddard made some bold connections between fire and the ideal quail habitat found in the Red Hills. His 1925–26 report argued there was a strong correlation in the longleaf forest between the presence of fire and the preferred habitat of bobwhite quail. The "open woodlands of some types," wrote Stoddard, "produce great quantities of favorite quail foods in the form of small legumes, such as butterfly and milk peas, tick-trefoils, bush clovers (*Lespedeza*), partridge pea, dwarf sumac and others."[35] His early work in the laboratory verified that quail favored the understory grasses and legumes, and in the wild he found them most commonly in burned over woodlands, not on lands kept in the rough. It was evident, then, that fire was not only a cultural necessity but also an ecological one, at least if a significant quail population was the goal. His report continued, "Some types of woodland in this region . . . have a tendency to grow up to jungle-like undergrowth, and unless controlled by late-winter burning the only alternative seems to be expensive brushing out by hand at intervals. Such mixed woodlands of pine and hardwoods . . . are too dense for quail."[36] If raising quail in the wild was the primary management goal on coastal plain lands, then the use of fire to maintain habitat was not only preferable, but a practical necessity. When the rough became too thick due to a policy of fire exclusion or simple oversight, Stoddard recommended opening the forest up with a mixture of low-level fire and manual hand labor, and he reported that "after such areas are opened up so that the sunlight reaches the soil, a luxuriant ground growth springs up, and this usually includes many plants furnishing valuable quail foods."[37]

Stoddard's interests in natural history extended further than a narrow focus on increasing quail numbers, however. Though he did not use contemporary ecological jargon like "vegetative succession" or "climax state," he did make some preliminary suggestions about the development of longleaf woodlands as opposed to other forest types:

> While the exact effect of fire on the many kinds of vegetation that contribute to the quail's food supply and cover can be determined only by observations over a long series of years, it has been possible to study the subject from the comparative standpoint on preserves in quail country of various types where the fire policy of each is known. Fire is unquestionably a controlling factor in determining the types of woodland in any given area in this region, as well as in the regulation of the ground vegetation.[38]

The natural ignition or cultural application of fire had been so frequent for so long that most plant species in the region had adapted to it and even become dependent on it, Stoddard suggested. This conclusion entered the public record as a policy recommendation, a result that landed him in the middle of a rhetorical firestorm. His recommendation to use fire as a strategy to restore quail numbers was perhaps the most significant, and certainly the most controversial, result of the quail study. But more than that, Stoddard quickly came to understand that the entire longleaf pine system required fire.

Trumpeting the use of fire ultimately helped to define Stoddard's legacy in the southeastern forests, but when his report filtered out of the South in 1925, arguing that fire benefited forest health in the longleaf region, the nation's foresters grew uneasy. This was not a good time to advertise the benefits of fire in southern pine forests. Government and industrial forestry organizations were in the middle of a decades-long attempt to secure administrative control over local resources, and practices like woodsburning were fast becoming anathema to good management. And the assumption that fire destroyed southern pine forests was not without some merit. After the massive industrial logging effort from 1880 to 1920, foresters had, in fact, witnessed large swaths of young second-growth forest swept over by fire.

This turn-of-the-century logging was a key turning point that brought national attention to the problem of cutover lands and fire protection in the South. Large timber companies turned their focus from the Great Lakes states to the South in the 1880s, where they found millions of acres of timberland still in the possession of state and federal governments. They also found southern lawmakers more than willing to hand it over. The first step toward opening the public domain to the timber industry was the repeal of the 1866 Southern Homestead Act, a Reconstruction-inspired piece of legislation that limited public land sales to eighty acres. The act was designed to offer small plots from the public domain to both freedmen and poor whites, but such Jeffersonian ideals did not last long after southern elites regained control of state and federal congressional seats. Stirred by the promise of northern capital, redeemed southern legislators pushed to repeal the act in 1876, and timber companies and other speculators soon proceeded to amass huge tracts of land and construct rail lines into the southern interior.

As it turned out, southern capital played almost as large a role as northern capital in the land grab, but one fact was clear: much of the southern landscape was falling into the possession of a handful of purchasers. According to historian Paul Gates, between 1880 and 1888, only 159 buyers purchased over 3.7 million acres of federal land in Louisiana, Mississippi, Alabama, Arkansas, and Florida alone. And these figures do not even consider the tens of millions of acres sold off by the states. After witnessing so many acres fall into so few hands, Congress halted sales of more than 160 acres in 1888, but the deed was done. Well-capitalized timber companies now held title to a large expanse of the southern coastal plain, and, with the help of local operators, proceeded to clear the longleaf forests on a scale previously unknown to the South. As forest resources dwindled, Forest Service chief William Greeley reported in 1925 that "the production of southern timber passed its peak in 1916 and the last great migration of American sawmills is under way across the Great Plains to the virgin forests of the Pacific Coast."[39]

In the wake of this industrial cut, lumber company officials, government planners, and local townsfolk had little idea of what to do with cutover lands. The problem was particularly acute in the coastal plain. Outside of the floodplains of coastal plain rivers and a few anomalous upland sections like the

Red Hills, the soils in much of the region were sandy, fast draining, and acidic—not at all ideal for farming. Early travelers called the coastal plain uplands the "pine barrens" for good reason. There were, of course, farmers spread throughout the region at the turn of the twentieth century, but they usually stuck close to the richer bottomlands that bordered rivers. Those who did venture into the sandy uplands engaged in diverse farming activities that utilized the region's open range.[40]

Nevertheless, converting cutover land to farms was a popular idea among planners. Some suggested selling farmsteads at discounted rates to returning World War I veterans. Others hoped to expand improved pasturage to make the region a national center for cattle production.[41] In the end, however, the various resettlement plans never took off. The most practical answer to the problem of cutover lands was reforestation. Many large tracts reverted to the public domain after deforestation due to unpaid taxes and were sold again to a second generation of lumber companies more interested in staying in the region beyond one cut. It became clear to these lumber companies and other local woodland owners that their enterprise would not last long without an intentional effort to reestablish the pine forests. Lumber company officials, civic leaders, and local landowners thus accepted the help of scientific foresters to institute modern methods of reforestation and forest protection.

The earliest foresters to enter the southern coastal plain saw limitless possibilities for the region, but the reality on the ground was less than impressive. Scientific forestry was practically nonexistent in the South, and administrative control of forest resources was piecemeal and had little influence over local activity. Some of the earliest calls for reform came from North Carolina, where by the 1890s the turpentine industry was ending a long run in the state, and timber industry loggers were already looking elsewhere for supplies of wood. W. W. Ashe, a prominent botanist and forester with the North Carolina Geological Survey, was one of the first officials to lobby for more state control of forestlands. As Ashe witnessed the timber industry's march across North Carolina's coastal plain, he became convinced that "the care and propagation of timber trees *en masse* becomes a feature of economic administration; and in all cases the inauguration of the policy of forest cultivation has emanated from the government."[42] Ashe was echoing the Progressive-era gospel of foresters such as Gifford Pinchot, and, like Pinchot and his followers, was much enamored with European models of scientific state forestry.

The practice and administration of French and German silviculture, in particular, was already a major influence on American foresters, and Ashe considered the South especially ripe for its importation. Old growth forests, which the European models considered inefficient and wasteful, were quickly disappearing, essentially leaving a clean slate ready for a uniform, scientific brand of forestry. The conceptual lens of European forestry was agricultural; that is, trees were a crop to be cultivated, rather than a resource to be extracted. Throughout the 1800s, Germany's state forestry structure, for example, had converted the nation's wild mixed forests into regulated single-species crops with predictable annual yields. Ashe and others hoped to implement a similar structure in the South. In a seminal 1894 article entitled "The Long Leaf Pine and Its Struggle for Existence," Ashe wrote that European nations such as Germany "have many years ago undertaken to place all their own forest lands under systematic management and at the same time supply, by means of their schools of forestry, the knowledge of these methods to private land-holders or to trained officers who may serve them."[43] As Ashe saw it, this model was an ideal solution for replacing southern forests.

The expulsion of fire from the range would be a central tenet of such a model, but as Stephen Pyne has argued, "Only when they were confronted with light burning as a defined alternative did American foresters invent the program that became know as systematic fire protection."[44] Widespread activities like woodsburning were not only damaging to the forests, officials like Ashe thought, but were also anathema to the ideals of scientific forestry. It represented the seemingly arbitrary decisions of locals who had no knowledge of botany, silviculture, or forest economics. Ashe believed "that sooner or later the present management or lack of management which has characterized all dealings with the barrens for the past 140 years, must be changed if the long leaf pine forests are to be made self-propagating," and that "the burnings of the present and future, if not soon discontinued, will mean the final extinction of the long leaf pine in the State."[45] Despite such early calls for systematic forest management, progress was slow. Social geographer Rupert Vance observed as late as 1932 that "the shift in timber from a mining to a cropping system of utilization has hardly made a start." Vance felt reforestation in the South depended on "the development of a national attitude toward forestry," but more than attitude, foresters needed knowledge about the mechanics of southern pine growth.[46]

There were few places to gain that knowledge. Following up on Gifford Pinchot's work as the forester on George Vanderbilt's Biltmore Estate in

North Carolina, German forester Carl Schenck established the South's first forestry school there in 1898, but its focus was more national than regional. The University of Georgia cobbled together the first forestry school with a southern focus in 1906, and Louisiana State University would not start the region's next forestry program until 1925. Within the state governments, Texas and Louisiana organized state forestry departments in 1915 and 1917, respectively, and Alabama, Georgia, and Mississippi followed suit between 1924 and 1928. These state departments formed at the behest of the various forest industry associations who hoped a state organizational structure might provide a means for fire protection, but they did little to advance scientific knowledge about the South's pine forests. The Forest Service made a major investment in southern forestry in 1921 with the establishment of a series of experiment stations, beginning with the Southern Forest Experiment Station in New Orleans.[47] The department line on fire, however, remained the same for many years. One experienced Forest Service official, Inman Eldredge, admitted in 1935 to a dogmatic ideology on fire within the Forest Service. He wrote that the service "rallied behind the premise that the only way to manage any forest anywhere in the United States, from Alaska to Florida, is to cast out fire, root, stem, and branch, now and forever. We have, with closed ranks, fiercely defended this sacred principle against all comers and under circumstances and any forester who questioned its universal application was suspected of treason or at least was considered a dangerous eccentric."[48] Even when foresters suspected fire to be useful—and some did—forestry's professional culture prevented open discussion.

The longleaf-dominated coastal plain represented a special problem that no one really understood. Because of its dependence on fire, and the political baggage fire carried, this environment did not easily lend itself to scientific consensus. The result was a fundamental disjuncture between ecological reality, scientific inclination, and administrative aspiration. One problem lay in a relative lack of experience among Forest Service personnel with fire-adapted forest ecosystems. Most Forest Service officials trained under the European model of forestry, a model in which forest fire was not even a variable; and most of their experience was in the Northeast and Great Lakes forests, where, due to large fuel loads after heavy logging, fires were more often raging conflagrations than low-level creepers.[49] And northern fires were less frequent anyway. For a state agency with influence over so many different forest types, a broadcast solution to the problem of fire was simple and would hopefully instill local populations with anti-fire doctrine. In the longleaf pine belt, though, routine fire was necessary not only to reduce fuel loads but also to facilitate longleaf regeneration, a fact that nobody quite understood at the time.

On this issue, the natural regeneration of what once stood, foresters were most flummoxed. Though it is impossible to generalize regionwide about a process as complex and localized as vegetative succession in the southern coastal plain, there are a few clues as to how foresters misinterpreted what they witnessed after the cutover. After an industrial cut on the upland coastal plain, less fire-tolerant pine species such as loblolly and slash—along with a host of hardwoods and shrubs—replaced the cutover longleaf. When trained foresters saw debris-fueled fires sweep through the region, killing these early successional pines, they assumed fire made *all* pine propagation impossible. As it turned out, they were wrong, at least if their goal was to regenerate longleaf woodlands. Many timber companies operated a high-grade cut; that is, they sought the oldest, straightest trees and left many smaller trees behind, effectively providing a seed stock for the next longleaf forest. In these disturbed longleaf landscapes, fire was necessary to expose the bare mineral soil needed for seeds to germinate, and to suppress the early successional pines and hardwoods, so the longleaf seed stock could establish its dominance. But without fire these seeds rarely had a chance. At the time, however, the mechanics of longleaf regeneration were largely a mystery.[50]

On many large tracts longleaf pine virtually disappeared, but where burning practices persisted under the system of free-range grazing, the second-growth longleaf forest made a comeback. Indeed, some government foresters noticed this phenomenon upon reflection. E. L. Demmon, of the Southern Forest Experiment Station, noted in 1935 that "many examples are to be found throughout the belt of excellent second-growth longleaf stands, which are known to have been burned over at frequent intervals over a long period of years."[51] Inman Eldredge also harbored suspicions about fire exclusion. He did not think a forester with five years experience in the South could "accept without reservation the premise that fire everywhere and under circumstances is an unmitigated evil. Many of them are convinced that in the reproduction of longleaf pine, fire is a valuable silvicultural tool."[52] The administrative momentum of the Forest Service's fire policy, however, stifled any constructive criticism from within. Another forester who came to the South around 1920, Austin Cary, revealed fifteen years later that "in the years during which I have been South, I have said as little as I could [about fire] in a public way. . . . For one thing, there was an official policy which one would not care to counter unnecessarily."[53] Foresters on the ground had little recourse but to remain silent about their suspicion that fire could actually aid tree growth.

It was a silence that carried tremendous environmental consequences. Through its destructive cutting, the timber industry deeply wounded the original longleaf forest, but the growing influence of scientific forestry—and the fire suppression it promoted—was the most powerful agent of ecological transformation in the southern coastal plain. Within a matter of years, large swaths of the longleaf range became a dense tangle of slash pines, loblolly pines, and hardwoods fighting for dominance and shading out the former vegetative composition of the region. Within this context of ecological change, fire suddenly became a very real problem. No longer would it creep along the forest floor, relatively easy to control while consuming native grasses and securing longleaf's preeminence; in this new ecological complex, fire would rush toward the forest canopy, consuming everything in its path and leaving destruction in its wake. Under these conditions, fire became impossible to manage. In their exclusion of fire, the new forestry establishment created an environment all the more vulnerable to a conflagration, which only reinforced the misapprehension that fire was the problem.

In the Red Hills, the landholding patterns of the quail preserves largely sheltered the region from the machinations of the timber industry and subsequent fire suppression. As a result, it was one of the few remaining pockets of fire-maintained longleaf pine woodlands in the South, which made it an ideal outdoor laboratory to study natural processes. This was an environmental space relatively free from the ecological changes of the region at large, and the woodland preserves of the Red Hills still operated under the older land-use patterns and practices. In addition, the landowners and tenants were of like mind in regard to fire. Elsewhere in the coastal plain, the longleaf forests were gone, local and national forces had a new vision for forest resources, new landholder goals began to close off the open range, and an increasingly complex system of railroads and highways began to compartmentalize the landscape in ways previously unknown. Even when some resource managers and local people saw its efficacy, fire became increasingly problematic. A fire that crossed property boundaries now had the potential to offend a neighbor whose land use eschewed fire, as well as destroy their property. These and other factors led to the desire of local civic leaders, corporate forestry interests, and scientific foresters for fire protection and suppression. The Red Hills quail preserves, however, functioned as one large unit; their focus on quail production and their close proximity to one another allowed them to continue burning as in the past. The environmental stability of Red Hills forestland, in other words, gave Stoddard a natural laboratory to observe the effects of fire under optimum conditions.

By the time Stoddard carried out his quail study in the 1920s, however, a stronger organizational structure allowed scientific forestry to penetrate further into the region. The American Forestry Association (AFA), with the cooperation of the Forest Service, state forestry commissions, and industry-oriented associations, carried out a series of anti-fire projects that reached deep into the southern woods. The most organized effort was the AFA's Southern Forestry Education campaign. Beginning in 1928 they sent scores of young foresters, known as the "Dixie Crusaders," into the South to preach against the sins of woodsburners. The brainchild of Ovid Butler, executive secretary of the AFA, the campaign focused on rural Georgia, Florida, and Mississippi, where the Dixie Crusaders sought "public enlightenment . . . among the rural people of the South who have followed for generations the custom of burning the woods annually," according the project's final report. The custom of burning the woods, "based upon ignorance of forest growth and forest values, was common throughout the entire South and was recognized by foresters as the most difficult barrier to forest regeneration and permanent forest management on over 200,000,000 acres of wild land in the South."[54] The general objective was to promote "a new and different picture of forests as a land crop," and to make southerners aware of the harm caused by fire to the forest industries, grazing conditions, wildlife, and scenic values.[55] In three years of activity, the crusaders toured the region in a fleet of trucks equipped with generators and motion picture projectors, and conducted more than 4,600 film showings, delivered more than 7,300 lectures, distributed 930,000 pieces of literature, and reached an estimated 2,679,000 people, white and black.[56] The message of the AFA and the Dixie Crusaders was diametrically opposed to what Stoddard discovered about his corner of the southern piney woods, partly because his corner was the diametrical opposite of the much of the cutover coastal plain. The longleaf pine woodlands remained, and fire posed little threat. Beyond these different environments, each party also arrived at different conclusions because they each began with a different set of desired environmental and economic objectives in the forest.

Stoddard did his best to ignore the AFA and the Forest Service. The cooperative nature of Stoddard's position gave him some autonomy in carrying out his study, but even a quasi-government agent was sure to catch some flak for espousing the beneficial role of fire. When *The Bobwhite Quail* hit the market in 1931, fire was by far the most common topic of conversation in land management circles, and few dared to speak of its propriety. It is a wonder *The Bobwhite Quail* saw the light of day as a government-sponsored monograph. Since both the Biological Survey and the Forest Service fell under the administration of the USDA, the manuscript ran through the Forest Service editorial office, and the foresters who saw it did not like what they read. As early as June 1926, the chief of the Forest Service, William Greeley, questioned the legitimacy of the Biological Survey in matters of forest management, in terms both biological and sylvic. He asked E. W. Nelson to rein in his charge, writing, "It is to be hoped that . . . Mr. Stoddard will be very guarded in the matter of fire and woods burning so as to guard as fully as possible against any possibility of the public misconstruing his statements to make it appear that the Federal Government advocates burning the woods in order to improve conditions for quail." On the contrary, Greeley contended that the "common practice of yearly burning the woods is effective in large measure in the depletion of game animals and birds. This is one of the standard reasons advanced by State and Federal foresters for preventing woods fires."[57] In other words, the Forest Service had ultimate authority in the woods, and it did not need a rogue federal agent instilling doubt in the minds of locals. Nelson was quick to defend Stoddard, making it clear that fire suppression "renders great areas absolutely worthless for quail."[58] But he also urged Stoddard to make plain "in your letters, talks, or publications" that the need for fire "is due wholly to local conditions and not for general application."[59]

To a certain extent, Nelson was right in his caution. The subtleties of burning depended almost entirely on local conditions, and locals were not always subtle in their practices. A scorching fire during the height of summer could do a great deal of damage to both wildlife and timber resources. But the Forest Service was not one for subtleties either; it had waged a strong and indiscriminate campaign against any sort of forest fire, and Stoddard's recommendations clearly threatened to challenge that orthodoxy. After his chapter on fire in *The Bobwhite Quail* went through five rounds of editing by Forest Service officials, Stoddard had to qualify his recommendations. He tempered the chapter with many remarks on the dangers of uncontrolled fire and the localized nature of his own study. He also conceded that the effects of frequent burning on flora and fauna were "a complex problem, one that would require years of careful research." The charge of his study was to increase quail, and "*they can at best be regarded only as a supplementary crop on timber and agricultural holdings*. For this reason we desire to have it distinctly understood that most of our remarks and recommendations in regard to the use and control of fire refer to lands owned and held *primarily for quail shooting*, and should not be used to embarrass the forester in his attempts to protect forest growth over the region at large."[60] Despite these concessions, Stoddard made it clear that where a landowner desired quail in the longleaf woodlands, fire was a necessity. Indeed, the chapter's opening line left little room for misinterpretation: "The bobwhite of the Southeastern United States was undoubtedly evolved in an environment that was always subject to occasional burning over."[61]

Stoddard was sure of his findings, but this experience with the Forest Service and the bureaucratic channels of government work made a lasting impact. He thought the editorial hedging allowed some foresters to take "sentences here and there favorable to their cause on numerous occasions that completely reversed the findings of the Quail Investigation," and that "there seems no limit to the lengths they will go to support their propaganda against fire."[62] Just as some practicing foresters found in the South, a policy that did not reflect environmental reality hamstrung Stoddard's ability to make public what he observed in the field. Shortly after the book's publication, he confided to one friend that "it is by no means as strong as I feel on the subject," and to another that "perhaps I met the foresters a little too far on their side, for fire is the one great, outstanding shaper of southern ecology, without which we would have a very different country and a vastly different (and less numerous and diversified) animal life."[63] After letting official reaction to the book simmer for a year, he predicted, in a letter to botanist Roland Harper, that he would "write a good deal more on the subject before I get through, for I am pretty well warmed up about it."[64] Despite his reservations about *The Bobwhite Quail*, it turned heads throughout the natural resource professions and helped to spark a thorough reconsideration of fire.

Stoddard followed up on his pledge to write more on the subject of fire, and immediately after the book's publication he was no longer much concerned with embarrassing foresters. He left the confines of the Biological Survey in 1930 and became a private consultant to hunting preserves throughout the South, where he hoped to spread the word about the beneficial role of fire. Public and private forestry interests, however, ratcheted up their calls for fire prevention, and New Deal programs like the Civilian Conservation Corps (CCC) eventually provided an apparatus for state and federal governments to place people and money in the field to administer protection units and fight fires. Stoddard became especially incensed about general appeals by forestry interests to use state and federal funds for outright fire suppression. The Clarke-McNary Act, a major piece of legislation in 1924 that gave states more authority in administering forest resources, vaguely hoped to split funding for fire protection down the middle between private landowners and government agencies. Not surprisingly, by 1932 "this plan has not so far worked out very satisfactorily," according to Fred Morrell, the Forest Service's representative on the CCC advisory council. In fact, "only in a few States are the private owners putting up more than a small percentage of the total amount needed."[65] Raising fire suppression funds in the South was a particularly tough problem. Burning in late winter was as routine as breaking the land in the spring; why would landowners pony up to prevent something so ingrained, especially when it was still not entirely clear that timber was a worthwhile cash crop?

As chair of the Technical Committee of the National Land Use Planning Committee, Morrell suggested a more concrete solution: "that the protection of forests from fire should be regarded as a public responsibility to be financed by general taxes."[66] Stoddard caught wind of the plan through Aldo Leopold, who was on Morrell's committee, and initiated one of his many campaigns to protect local prerogatives in the maintenance of forest land. "At present," wrote Stoddard,

> there is in this region in excess of a million acres owned primarily for quail shooting, and to a lesser extent for the production of wild turkeys, deer and other game, and this acreage has been rapidly increasing. . . . In addition the quail shooting rights are leased on a much larger acreage, while public shooting of quail is undoubtedly one of the most important forms of recreation; so important that the matter should be considered in any "land use" discussion.[67]

Thus began a treatise on fire that Stoddard repeated in various forms for many years to come. He first made it clear that in a rapidly expanding industrial world, game animals could no longer be taken for granted—the intentional management of game resources was essential to meet the public demand for them. By invoking the public's right to game animals, he co-opted the rhetoric of other natural resource managers, thus providing an economic justification for game management that was anything but self-evident.[68] Such public demand meant that any attempts at government land-use planning had to recognize the distinctiveness of local and regional environments, and take locally established management techniques into account. In the longleaf pine belt, those techniques included the use of fire. In his letter to Morrell, Stoddard continued with a focus on game animals, but expanded his view to make clear that fire was essential to most types of land use in the longleaf woodlands, while also offering a glimpse into the broader ecological role of fire that he would embrace more fully a few years later:

> Controlled burning of these pinelands at intervals averaging about every alternate year is practically essential to maintaining them in condition for quail . . . and further studies since [our] Investigation show that the use of fire is even more essential than at first believed. Never-the-less a vast amount of pine timber can and is being produced on quail preserves, where the timber is much more apt to be held until fully mature than on surrounding lands. Taxes to be devoted largely to fire exclusion, under the mistaken impression that fire is a curse no matter when or how used, is obviously unfair to owners of such acreage, though public funds that could be used to aid them in controlling necessary burning, as it is by others in fire exclusion, would doubtless aid this very worthy land use movement.
>
> In addition, there is no uniformity of opinion on the part of land owners as to desirability of absolute fire exclusion . . . over long-leaf pinelands in particular, and in other pinelands after they reach a certain age, for there are an ever increasing number of students of Southeastern ecology who consider properly used fire a valuable silvicultural agency. . . . If public funds are to be made available for "protection" of Southeastern pinelands, they should be made available to those who favor wise use of fire to prevent their holdings from growing up to deciduous jungle unsuited to quail and wild turkeys, and with a tremendous fire hazard.[69]

Morrell's plan for a general tax never passed, but all of this high-level government interest in local policy made Stoddard apprehensive about the legality of burning. Laws prohibiting arsonous woodsburning were long on the books (and rarely enforced), but the vehement tone of the post-cutover campaigns pointed toward outright criminalization of any type of woodsburning. The many appeals for fire protection and suppression caused Stoddard to worry that "landowners would not be permitted to do necessary burning, even when every precaution was taken to confine the fires to their own property, and when they had very definite ideas as to just what they wanted to accomplish."[70] His long-term goal was to reestablish fire across the coastal plain and study its behavior and effects under controlled scientific experiments, but Stoddard's immediate concern in the early 1930s was over the legal right to burn. Fortunately, he was not alone.

As Stoddard became surer of his ground on fire, he began to make contact with a small number of isolated individuals who were drawing their own conclusions about fire's role in the longleaf pine. As he told Morrell, there were others in the region concerned with forest ecology, and he would form with them a shaky coalition of like minds but manifold motivations. Many of Stoddard's preliminary conclusions about fire echoed those of the South's eminent botanist, Roland Harper, a surly spokesman for native southern flora who theorized about longleaf development many years earlier. Harper had many years of experience traversing the southern coastal plain by the time Stoddard came along. He completed his Columbia University doctoral dissertation in 1905 on the Altamaha Grit region, a unique expanse of subsurface sandstone in south Georgia, and worked intermittently for the geological surveys of Florida, Alabama, and Georgia over the next several decades. Those jobs sent Harper exploring throughout the southeast, and the influence of fire on the botanical composition of the coastal plain became one of his principal interests. In his 1914 report to the Florida Geological Survey, he wrote that in the longleaf forests "it is evident that fire is a part of Nature's program."[71] Harper noted the longleaf pine's resistance to fire and its comparative advantage in such a landscape to other pine species. He also speculated that lightning-caused fires once burned unmolested over large areas of the coastal plain, but "now they are mostly of human origin. Although fires are more numerous at the present time than they were originally, the area over which each one can spread is limited by roads, clearings, and other artificial barriers, so that the frequency of fire at any one point may not have changed much."[72] Stoddard became well acquainted with Harper's writings on fire, and the two began a long-lasting correspondence early in the quail study. Harper had little patience for the new forestry that infiltrated his region from the North, but as an academic botanist he had little real influence on forest policy. He thought Stoddard did, however, and encouraged him repeatedly to take on "the forestry crowd [that] is still preaching the same old nonsense about the damage done by fires . . . in a desperate effort to save their traditions."[73]

Another important voice in the fire controversy was S. W. Greene, a USDA Animal Industry Division field agent at the McNeill Experiment Station in Mississippi. Stoddard met Greene in Jacksonville at the 1929 meeting of the American Forestry Association, and they immediately struck up a correspondence. Greene was in the midst of cattle nutrition studies at McNeill, where preliminary results suggested burned over lands yielded more nutritious cattle feed than lands kept in the rough. Greene presented his work in Jacksonville, and despite Stoddard's observation that "there was little in the program to assist in my work," he thought "the paper of Greene was really very valuable."[74] Later that year, on an inspection trip for the Biological Survey, Stoddard visited McNeill and toured the property with Greene. He was most impressed with Greene's work, especially his experiments testing the efficacy of controlled fire and grazing on forestry holdings. He also considered McNeill an ideal place "to carry on studies as to the possibility of increasing and maintaining quail on forestry-grazing land," but that idea never materialized.[75] Stoddard was primarily relieved to know someone else—from within a government department, no less—who exposed the practical benefits of fire. Greene became an important ally of Stoddard's as they sought to challenge the rhetoric of the fire suppressionists.

Other allies came from within the ranks of forestry itself. In fact, despite a vociferous defense of fire suppression among most foresters, experience in the woods led some forest managers to at least privately question the motives of agencies like the Forest Service. One of those managers was Henry Hardtner. Though not a trained forester, as head of the Urania Lumber Company in Urania, Louisiana, Hardtner was one of the most forward-looking lumbermen in the South. As early as 1918 he suggested "'controlled burnings' should be practiced in every forest as an aid to successful forestry."[76] The year before, Hardtner offered the use of his holdings at Urania to H. H. Chapman, an influential forester at Yale. As one of the first scientific foresters conducting research in the region, Chapman carried out a host of experiments on the relation of longleaf reproduction and growth to fire. His publications from these studies, especially "Factors Determining Natural Reproduction of Longleaf Pine on Cut-Over Lands" in 1926, and "Is The Longleaf Type a Climax?" in 1932, were landmark articles.[77] In them, Chapman addressed the various methods of reforestation found in the southern coastal plain, and argued that using fire to encourage the natural regeneration of longleaf pine was most desirable. Other restocking plans, like planting nursery seedlings of loblolly and slash pine, would "work against indications so clearly shown that in the struggle between these species the Longleaf pine is the natural survivor"; such a plan would expose the "lesser" species to "conditions abnormal to them," namely fire.[78] His studies at Urania revealed that the most effective method of longleaf regeneration was to burn the range shortly before seed fall, and again every few years after germination. This was controversial fodder for the forestry profession, but, as his tenuous relationship with Stoddard and Greene reveals, Chapman was anything but an apologist for traditional southern woodsburning.

The master link between these scientific fire starters was the environment itself—the fire-dependent longleaf-grassland system. They each found value in the forest, whether it was in forage, wildlife, or the woody fibers of longleaf pines; and they each found fire to be useful in exploiting that value. But as these early renegades began to form their fire coalition, their differing expectations about what this environment was supposed to produce led to little consensus about how to proceed on the ground level. Their internal spats reveal the developmental nature of a new field of science that bridges many biological disciplines. Fire ecology did not respect the disciplinary boundaries of wildlife management, forestry, range management, or agriculture, and those practitioners inclined to harness fire hoped it would work expressly for their interests, which created a certain amount of volatility even within such an agreeable group.

S. W. Greene was the most vocal of the bunch. He fired off a volley of letters to the AFA's publication, *American Forests*, in 1931, lambasting the "travesties in truth published in AMERICAN FORESTS for a number of years."[79] His caustic tone got the attention of Ovid Butler, executive secretary of the AFA and editor of *American Forests*, who, to his credit, allowed Greene a voice in the October 1931 issue of the magazine. Greene's article, "The Forests That Fire Made," was probably the first article for popular publication that redressed the South's anti-fire propaganda. In it he drew on his research to not only make the case for tightly "controlled" burning, as had Stoddard, but to also defend the traditional types of burning that foresters had come to call "uncontrolled" and destructive. He wrote, "Thoughtful men of the southern piney woods have stood calmly by and made . . . a study for generations without knowledge that the study of the relation between trees and their environment had such a scientific classification as ecology. Their conclusions were not set down as scientific treatises but were passed along under the more common name of woodcraft and the lore of the woods told them when, why and how to use fire."[80] This was just the type of burning the AFA was out to stop, and Greene went on to accuse the fire suppressionists with transforming the coastal plain's forested landscape. He explained that longleaf pines established their historical dominance in the upland coastal plain because their resistance to fire gave them an advantage over other pine species, and "that other species crowded out longleaf only in the absence of annual grass fires."[81]

Stoddard was enthusiastic in his support of Greene's efforts, but Hardtner and Chapman had reservations. They agreed that longleaf forests developed in concert with fire, but were in no hurry to broadcast their findings to the public. Before publication of the article, Chapman wrote Greene, saying, "Unless we can properly control and hedge the propaganda regarding fire so that an average southern white farmer can understand what we are talking about . . . I would prefer to have very little said about the use of fire. . . . Personally I have no intention of bursting into print until the conclusion of certain experiments."[82] As previously noted, Chapman's longleaf research did reach print, and it was indeed groundbreaking. His academic outlets, however, were not likely to reach many southerners. In Chapman's mind, Greene was engaging in a dangerous campaign. Discussions on burning among the experts were all well and good, but Chapman believed Greene wanted to broadcast generalizations about fire that would reinforce generations of "miscreant" behavior in the South. Chapman told Hardtner that he would be "exceedingly sorry if Mr. Greene or any other agency became responsible through their statements for the publication of misleading, false, and mischievous statements regarding the promiscuous use of fire."[83]

Greene's *American Forests* piece did, in fact, lack the usual cautionary statements about uncontrolled fire, and even implied that local woodsburners were the true experts in the field, a proposition that made Chapman recoil. As a scientist firmly rooted in both the academic and industrial forestry establishment, Chapman did not want local populations to suspect they had been right all along. The coming of industrial forestry to the South meant a tightening control of traditional local activity in the woods. Chapman felt "much sympathy with the state foresters and the American Forestry Association in their warfare against the habit of indiscriminate and uncontrolled burning," and a public airing of any counter claims threatened to "cause confusion and loss in their effort to create a fire consciousness in the minds of the southern people."[84] There is little doubt that southerners had a fire consciousness, just not in the way Chapman wanted. He preferred a scientifically considered rationale for burning, and strict supervision and application of fire by trained experts, not a seemingly random ignition based on the whims of a local rural dweller. Indeed, Stoddard and Greene were interested in the more orderly application of fire as well, only from a different perspective.

The forest understory was foremost in the minds of Stoddard and Greene, regardless of the forest type it harbored, and their approach to burning more closely resembled that of traditional woodsburners. Stoddard attempted to ease Chapman's mind about burning in different forest types, writing in a friendly but pointed letter that "we do a great deal of burning both of loblolly and long-leaf lands. . . . We have found that if the burning is carried on carefully at night when the dew is falling we can run fires through loblolly after it is a few years of age with little damage to the trees. . . . We have found that there are all kinds of fire and that a person can use any kind he desires, from extremely light to very severe."[85] This raises an important distinction in Stoddard's experience with fire. First, he recognized that fire suppression would not only transform the longleaf ecosystem; it would transform a way of life as well. As he told Ovid Butler, "The open pine forests of the Southeast have persisted for ages and are a most pleasant environment for man and many forms of wild life. There is absolutely no doubt in my mind that fire was a major factor in molding this environment. . . . So why go off half cocked, assuming that the inhabitants of the region are a bunch of ignoramuses, and try to force a complete reversal of custom before the facts in the case are known."[86] Second, as a wildlife manager, he was concerned with more than trees. He understood that fire could take on certain characteristics depending on environmental conditions, and a low-level fire in a loblolly, shortleaf, or mixed forest could create an understory habitat beneficial to a variety of wildlife and still not harm the trees. He, too, was interested in creating a science of fire, but it was one based in what we would call today applied ecology, not forestry.

The 1935 meeting of the Society of American Foresters was the real debut for controlled fire. Stoddard, Hardtner, Chapman, and Greene, along with longtime southern foresters Austin Cary, Inman Eldridge, and Elwood Demmon, as well as Ovid Butler, all put aside their differences to hold a panel on the beneficial role of fire that the audience largely greeted with acceptance. As happened in other conservation circles, longtime assumptions were being turned on their head in the forestry profession as well. The panel's content largely echoed what southern foresters had already said in more hushed tones, but this was the first public discussion among southern resource managers addressing the use, rather than the abuse, of fire. Some participants, like Henry Hardtner, maintained a latent concern "of the public not understanding our methods or plans," but he conceded that "the people who live in the forest . . . are forest minded and are actually acquainted with the very problems" of controlled fire.[87] Some attendees, however, were less optimistic. Forest Service official Roy Headley, though he had "much in common with many of the papers," offered a "good natured protest over the one-sided nature of the program. If any one is looking for evidence of censorship . . . he need look no further than the make up of the afternoon's program." In response to Headley, Stoddard's scrappy young assistant, Ed Komarek—whom we will get to know well in coming chapters—replied, "The whole program up to this time has been one-sided. This is the first time that censorship on the subject has been removed and we have been told the facts."[88] The panel was indeed groundbreaking for those who felt vilified by the fire suppressionists. When Stoddard returned from the conference, he wrote to Aldo Leopold, expressing a deep sense of relief: "If you are interested in the Southeastern fire question, you will find the papers and discussions of the last afternoon well worth careful perusal, as much of a revolutionary nature was brought out. For the first time in ten years I had a feeling that perhaps I was not a 'public enemy' after all."[89]

Despite this reprieve, Stoddard continued to debate government and private foresters over the use of fire for many years to come. Only a year after the panel, in fact, the Forest Service released a pamphlet entitled "Woods Burning in the South," which made Stoddard feel somewhat betrayed. He felt it was an "unqualified condemnation of fire use," in which some readers "might think [fire proponents] were compared to the boll weevil, malarial germ, and the cattle tick," three biological agents of destruction the government was desperately trying to eradicate.[90] There was obviously much work to be done. Federal, state, and private foresters continued to debate the use of fire for years (and still do today), but despite their doubts, at least the Forest Service and its constituency began to recognize controlled fire as worthy of debate.

How to apply fire in the southern woodlands, however, was another matter altogether. In fact, one of Stoddard's most important contributions to the fire debate was to develop a methodology for application. That a method needed to be developed at all reveals the knowledge gap between foresters practicing in the South and the local farmers and woodspeople who had lived in the region for generations. Stoddard wrote in 1935 that "in recommending liberal use of fire in maintaining quail ground, we are not suggesting anything *new*, for burning has long been customary in the Southeast."[91] Despite being an accepted practice by locals in the region, Stoddard's effort to legitimate his stance in the wider professional world required that he distance himself from older forms of woodsburning. When he set out to educate land managers on how to apply fire in the southern coastal plain, then, he hoped to negotiate a middle ground between the anti-fire forester and the fire-loving local. On its surface, Stoddard's system was practical, as well as simple, but industrial foresters did not come around easily. Even after organizations like the Forest Service came to agree that fire helped control the rough and aided in longleaf regeneration, there remained a deep apprehension about lighting up. For example, Stoddard told Roland Harper of one Forest Service official who admitted at a forestry meeting in Thomasville that "'controlled burning had to be carried on for upland game . . . but nobody knows how to do it but Stoddard.'" At this suggestion, Stoddard quietly quipped, "Most countrymen know more about controlled burning than he does about forestry."[92]

Again, though, Stoddard practiced land management within the context of a new political and environmental landscape, and the pressure from government agencies and state legislatures to make the use of fire coherent to a bureaucratic world led him to help codify the once free-wheeling practice of burning into a more rigorously applied science. In *The Bobwhite Quail*, and later in a series of more-detailed pamphlets, he encouraged land managers to subdivide the range into units, using where feasible roads and natural buffers, and where necessary plowed strips as fire breaks, "to be safely burned out as often as proves necessary without endangering neighboring units."[93] Whereas locals once set fire to a block of woods and let it burn until it simply went out, Stoddard helped to create a grid of well-defined fire units, making clear the distinction between controlled and uncontrolled fire. With the units in place, landowners or managers could implement a two-year rotation wherein they burned every other block yearly, thus leaving plenty of wildlife cover on the unburned blocks.

With their circumspect acceptance of burning as a legitimate management tool, the Forest Service also began conducting research on fire at the Southern Forest Experiment Station in New Orleans, but theirs was a relatively narrow focus on fire's affect on pine tree growth.[94] Stoddard, on the other hand, turned his energies to study what would become the consuming interest of forest and fire ecologists later in the twentieth century: how the frequency and seasonal timing of fire affected particular vegetative associations in the longleaf-grassland understory. This highly academic enterprise began as practical observation narrowly focused on quail food plants. The dissemination of partridge pea, for example, required close observation before randomly applying fire. An early fire might cause premature germination, making any new growth susceptible to late freezes. On the other hand, burning later in the spring could kill any young plants that already germinated in the rough. Stoddard advised that "good management calls for the burning of lands known to be seeded to [partridge pea] *after the danger from late frosts is largely past, and before growth has started.*"[95] Most other native legumes, however, were well adapted to fire throughout the late winter and spring months. Stoddard came to recognize the forest understory's relation to fire as a highly complex set of reactions that changed yearly depending on temperature and moisture levels, and he continued to argue that only close observation of seasonal growth could determine when to apply fire. Again, such complexity proved frustrating to government officials. The closest Stoddard came to a general burning formula was this pithy declaration: "The burning should be conducted when judgment and experience indicate that the greatest good and the least harm will be done."[96] He continued to publicly advocate for controlled burning and further fire research for many years, but when given a choice "between late winter burning of the kind long carried on by the natives of the region (who know more about fire use than usually given credit for) and the total fire exclusion policy so strongly advocated by some," he would choose the former every time.[97] He could not predict all of the environmental reactions to fire, but he knew it had to be applied. As he told one Forest Service official, tongue in cheek, "The animals know what is good for them, and as they flock to burns when available, we give them burns to go to."[98] In other words, the exigencies of the longleaf pine woodlands required practical application before full ecological discovery. The only way to figure out how this environment worked was to work with it.[99]

The 1935 meeting of the Society of American Foresters was by no means the end of the fire debate. Indeed, it was just getting started. Stoddard and others continued to work hard to accumulate facts and verify hypotheses, but the story of southern fire was about more than testable scientific knowledge. It was about a long tradition of setting fire to the coastal plain piney woods, and a sudden environmental transformation that made that tradition difficult under the best of circumstances. The traditional practice of burning the woods made an invaluable contribution to what eventually became accepted scientific knowledge about the longleaf-grassland environment. But in such a modern, fragmented landscape, fire had to be contained and controlled. The wave of scientific and industrial foresters who followed the industrial cut saw to that; not only had they attempted to suppress the use of fire, but they helped to create an environment that prevented its use.

Figure 9. Stoddard night burning on Sherwood Plantation, 1941. Stoddard not only argued for the ecological benefits of fire; he also developed many of the early protocols for using it. Spot burning at night, as Stoddard is doing here, was useful in dry conditions because the flame gathers less momentum, and the higher humidity levels after sundown help keep the fire under control. Courtesy of Leon and Julie Neel.

In this context, the quail preserves of the Red Hills were all the more important. They were one of the few spaces untouched by the transformation. Stoddard set to work in 1924 in an ecological bubble, insulated from both the world of professional forestry and the southern coastal plain's ragged environments. After his publications reached a wider audience, thus becoming one of the catalysts for a larger fire debate, he began to merge tradition with science. Proper application of fire, to Stoddard, was more than simply lighting a match in the old way, or developing a rigid set of rules and regulations in the new way. It was what he came to understand as an artful process, a system with predictable outcomes based on close observation of environmental conditions and experience. With his background in practical woodsmanship and his understanding of modern science, Stoddard translated for government and scientific professionals the fire-related activities of local people. It would not be his last time in the role of moderator.

CHAPTER 4

Stalking Wildlife Management

Herbert Stoddard's struggle with foresters over the use of fire demanded much of his attention before and after publishing *The Bobwhite Quail*, but his principal interest during these years was to carve out a niche for wildlife management as a professional field. He was much more interested in sorting out the complex interactions of wildlife, plant life, and human land use than getting caught up in the convoluted world of forestry policy. And no area in wildlife management needed more work than predator-prey relations. In contrast to fire, however, Stoddard did not consider the local perspective on predators a good place to start. He thought of local—and even regional and national—attitudes toward predator control as a mixed bag of uninformed myth and indiscriminate killing that had little to do with ecological reality.

Stoddard was far from the first to enter the fray on the issue of predators and prey. As many environmental historians have shown, national debates between predator eradicationists and protectionists raged in the decades leading up to the quail study. Sportsman groups and livestock interests multiplied rapidly in the early twentieth century and put a hard press on the federal government to protect their interests from the threat of predators large and small. As early as 1905, the Biological Survey participated with the Forest Service to trap wolves on public land throughout the West, and within a few years developed effective poison formulas and distributed them to cattle ranchers. Congress ponied up more funds in the early 1900s, but the demand for help quickly outpaced funding, leading the survey to enter into cooperative agreements with the cattle industry. From these agreements came a steady stream of funding and the establishment of the Biological Survey's Division of Predatory Animal and Rodent Control in the mid-1920s. According to Thomas Dunlap, "Scientific studies and the conservation of wildlife became less important [to predator specialists] than return on money spent and a high kill of 'varmints.'"[1]

Opposing such control measures was a growing group of scientists and naturalists who were concerned about the ecological and moral implications of predator eradication, as well as the loss to science that such programs meant. The American Society of Mammalogists openly criticized the Biological Survey in 1924 for exterminating local wolf and coyote populations in the West before conducting any substantial scientific research. Within the survey, a handful of scientists, such as Waldo McAtee and Albert K. Fisher, busily studied the food habits of birds of prey in an effort to stem their slaughter by farmers and sportsmen. In the same decade, biologists viewed the decimation of large predators in Arizona's Kaibab National Forest and the subsequent irruption and destruction of the local deer herd as a particularly stark example of misguided eradication efforts, making it clear to many just how little was known about predator-prey relations.[2]

Environmental historians have viewed this reconsideration of predator and prey as part of an American enlightenment in environmental thought. As ideas about balance and interconnectedness filtered into the mainstream through the discipline of ecology, many Americans began to see nature through the lens of any number of metaphors: webs, organisms, pyramids, communities, systems, even machines. Over the past few decades, the ecology of disturbance has raised questions about the accuracy of some of these metaphors, but they continue to carry a great deal of meaning nonetheless. And for good reason. They can be powerful representations for how environmental components might relate, and are especially useful in considering relationships between animal species that eat one another. Though ecology as a discrete discipline came to prominence mostly through the work of botanists and zoologists during the decades around the turn of the century, the rise of wildlife management during the interwar period had a great deal to do with making ecological ideas the basis for an ethical relationship with the land and its critters. Aldo Leopold wrote most elegantly about this new way of viewing the land, and scientists such as Stoddard, Paul Errington, Waldo McAtee, Charles Elton, and Leopold himself conducted much of the field science that would help prop it up. Their insistence that predators were an integral and necessary part of ecological processes was central to this ethical shift.[3]

But within this small scientific community, there was little consensus on how to implement research into a system of conservation-oriented land management. The science itself was rarely clear about how predators fit into abstract ecological models, and increasingly complex ecological thinking did not always lead to a biocentric respect for all predators and their ecological roles. What's missing in our examination of this scientific and attitudinal shift is attention to the peculiarities of regional environments and the ways in which local people interacted with them. The social context of environmental space was a critical influence on how scientists and land managers came to understand particular ecologies, and the different ways this convergence happened often led to earnest debate over the application of new ecological ideas to land management.[4]

Stoddard's work on predator-prey relations developed in association with a core group of early wildlife scientists, and the accessible ecologies and working conditions of each of these scientists helped to shape their contrasting approaches to the question. I am particularly interested in comparing the scientific work of Herbert Stoddard and Paul Errington, two early wildlife scientists who shared many things in common: they had similar backgrounds as hunter-naturalists; they helped to shape wildlife management into a tenable and important profession; and, perhaps most importantly, their research on the bobwhite quail and its predators helped to undergird the new environmental ethic.[5] Their scientific approaches, however, developed from particular places and had different goals, which led to friendly, but strong, disagreements on the specifics of predator-prey relations. Largely because of Stoddard's contacts in the region, Errington's quail research was in the countryside around Prairie du Sac, Wisconsin, an agricultural landscape on the northern edge of the quail range. The extreme winters of the upper Midwest made life hard on the bobwhite quail, especially in comparison with the slight seasonal variation in the southern coastal plain. Stoddard, on the other hand, had to factor in another type of extremity that Errington did not need to consider: the voracious human predator. Quail hunters were much more abundant in the Red Hills than in Errington's research area and had a much larger impact on quail populations. Because of these environmental and cultural differences, Stoddard and Errington went about their research in different ways and came to different conclusions about the role of predators. For a time, they even appeared to represent two competing perspectives about the value and purpose of wildlife management. They would come to recognize that their apparent differences really boiled down to differences in regional ecology and the human animal, but the remnants of their dispute—one between the practice of land management and the modeling of ecological science—came to typify some of the early growing pains of wildlife management, especially in the organization of the field's first professional group, the Wildlife Society. No facet of wildlife management better illustrates these early debates than the scientific problems of predator and prey.[6]

Much as they did for Stoddard's study of fire ecology, the Red Hills quail preserves offered him a vast land base on which to study the interactions of predators and prey, and space to implement management strategies to manipulate those interactions. And as we have already seen, this was more a cultural landscape than it was a wilderness; people were directly engaged with the environment in a variety of ways, making it impossible to study ecological phenomena in isolation from the activities of people. The study of predators was especially difficult because of old prejudices against many of them. Quail enthusiasts and locals alike placed few limits on what they considered to be quail predators. Diamondback rattlesnakes, red and gray foxes, opossums, raccoons, skunks, cotton rats, hawks of all species, and a wide range of other "vermin" were fair game. Some traditional activities, like the night hunting of raccoons and possums, Stoddard thought legitimate and effective predator control, even if that was not always their purpose. Others, like the indiscriminate killing of snakes and hawks long encouraged by preserve owners, he did not. Because of such complex interactions between people and animals, Stoddard's work became more than just a study of ecology; it also became a study of human ecology.

One example representative of the general attitudes toward predators was a landowner-sponsored contest for tenants that predated the rattlesnake roundups that commonly occurred later in the twentieth-century South. Longstanding apprehension over rattlesnakes, among both landowners and tenants, partly fueled the contest, but the sponsors also thought a recent increase in rattler populations responsible for diminishing quail numbers.[7] Beginning in the summer of 1922, at least five preserve owners signed a contract pledging to pay tenants one dollar for each diamondback rattlesnake killed within their property boundaries, as well as end-of-the-year cash prizes of fifty, thirty, and twenty dollars to tenants who killed the most. In the first year, tenants turned in a total of 631 rattlers.[8] When shown the tally from the next year's contest on his first day in the Red Hills, Stoddard was alarmed that "the grand total for the year's campaign amounted to over one thousand!" He was clearly annoyed about the myopic approach of the contest, but he was also disappointed about a lost opportunity to learn something of rattlesnake relations with quail: "Unfortunately no food study or statistical work of any kind was done."[9] Though preserve owners discontinued the contest in 1924, they continued to pay for individual snakes, so Stoddard set up a study of his own; it was "agreed that the managers of the various estates would bring me all snakes killed during the summer . . . for postmortem examination to find if they were feeding on quail or quail eggs."[10] Stoddard proceeded similarly with other predatory species, compiling a core data set not only to determine which predators actually preyed on quail and which ones did not but also to place quail in the broader ecological context of predator-prey relations, and thus to establish effective environmental control measures that discouraged the indiscriminate killing of predator species. *The Bobwhite Quail*, then, was one of the early studies to make recommendations on predators based on thorough scientific research. Though it focused on the bobwhite quail and was not a full study of animal population ecology, the study began to reveal the intricate relationships between animal populations and their environments.

Stoddard began the study just as national interest surfaced in game management, and disasters like the Kaibab made apparent the need for scientific research to serve as a guide to control. His research had four primary audiences: large preserve owners, government policy makers, academic biologists, and farmers—a grouping of interests generally bound only by bullheadedness. A fundamental problem for these interested parties was a general lack of consensus on what was predator and what was prey. Attendees at the 1925 American Game Conference, an assembly of assorted wildlife groups, attempted to rectify the problem. They enlisted Albert K. Fisher, a veteran investigator in the Biological Survey, to head a committee to identify those birds and mammals classed as "vermin." Fisher's specialty was birds of prey, and his bulletin, *The Hawks and Owls of the United States in Their Relation to Agriculture*, was among the first to suggest that farmers and hunters had little to fear from most raptor species.[11] Fisher's mandate from the game conference was to "make a list of vermin that will be a menace to game in all places at all times," but he quickly understood the futility of that charge.[12] Game animals to some were pests to others, and vermin in one region provided valuable agricultural pest control in another. Simply listing predators as "good" or "bad" would miss the complexity of local attitudes and environmental conditions.

Fisher consulted Stoddard, hoping to gauge the feasibility of his assignment, and to get a sense of predator control measures on the southeastern quail preserves. How did Stoddard deal, for instance, with species that preyed on a targeted game animal, but were also "of great service in removing more potent enemies?"[13] Stoddard almost pitied Fisher his task, and cautioned that a simple list was no way to approach the problem. Populations of so-called vermin species were interconnected not only with one another but also with game animals and habitat. He advised Fisher that "the question of course is such a complicated one that the listing of principal enemies of the quail in this region in the order of destructiveness can at best be but a rough approximation." He could list most quail enemies by this time, but the complexity of population dynamics made applicable control a difficult matter. Exterminate the gray fox, and cotton rat populations might explode. Control cotton rats, and their predators might turn to quail. This was a difficult business, and Stoddard himself was only in the beginning stages of discovery. He could, however, express his regret about the nature of the problem: "Personally, it is very distasteful to me to have any of our native wild creatures killed as vermin. . . . I sincerely hope that education will develop a more tolerant spirit among hunters and farmers towards predatory creatures which careful study shows are doing them no material damage."[14] Like others in his wildlife management cohort, Stoddard also brought core values to the table. But despite his moral inclination to leave predator and prey their own devices, he rarely considered the possibility, at least on the southeastern quail preserves. He hoped, instead, to use an ecological approach that incorporated humans as active participants in the dynamic relationships between predator, prey, and habitat. Indeed, he would eventually come to see a sort of symmetry between human and animal predators.

As Stoddard's study progressed and he sharpened his thoughts on predators, game management began to blossom as a professional field. A defining moment came in October 1928 when Stoddard joined with Aldo Leopold to guide the game fellowship program of the Sporting Arms and Ammunition Manufacturers Institute (SAAMI). A group with a vested interest in the nation's supply of game animals, SAAMI hoped to verify anecdotal evidence about diminishing game populations and reverse the trend through scientific study and restoration. Leopold began work for the institute in early 1928, and he was in the midst of a landmark game survey in several midwestern states when they brought in the Biological Survey to help construct a fellowship program for graduate students interested in game research.[15] Stoddard was the logical choice to represent the Biological Survey. His four years leading the CQI made him one of the only scientists in the nation experienced in the design and implementation of the new research in wildlife management. Leopold and Stoddard's charge was to make cooperative agreements with several universities throughout the United States, and identify suitable graduate candidates to carry out game-specific research based on the CQI model. Their collaboration gave birth to modern wildlife management.

Leopold and Stoddard first met in Cincinnati, Ohio, and were immediately struck by each other. As Stoddard remembered their first meeting, and the subsequent three-week trip through Ohio, Michigan, Illinois, Iowa, Wisconsin, and Arizona,

> I could see at once that we were kindred spirits. We shared almost identical interests, from ornithology and game-bird life-history studies to hunting and outdoor life. I was impressed by his mental capacity, his tremendous enthusiasms, and his high ethics and ideals. He had a stimulating personality, and in conversation he was able to draw out the thoughts of others, as well as freely sharing the depths of his own brilliant mind. He would think deeply and quietly a few moments, marshaling his thoughts in logical sequence, and then express them clearly, forcefully, and eloquently. Later I was to find him an ideal chairman for any sort of conservation meeting because of his extraordinary ability to grasp the basic aspects of a situation, discarding unimportant side issues and argumentative froth and presenting the fundamental points in a few choice phrases.[16]

Their initial time together not only sparked what Stoddard termed "one of the finest and most stimulating friendships and associations of my lifetime," but it also marked the ascendancy of wildlife management into a field of its own.[17] Scientists in other subdisciplines like ornithology and zoology certainly studied wildlife, but as a conservation-oriented science, wildlife management was fundamentally different. As Stoddard and Leopold devised it, the science of wildlife management carried with it an ecological perspective. That is, it would not study a wildlife species in isolation from its environment, and it would focus explicitly on maintaining and restoring habitat. Wildlife management, in other words, required land management.

On that first trip, Leopold and Stoddard made important contacts with university biologists across the Midwest, and they eventually identified graduate fellows at four universities: Ralph King at the University of Minnesota, Dave Gorsuch at the University of Arizona, Ralph Yeatter at the University of Michigan, and Paul Errington at the University of Wisconsin. King, Gorsuch, and Yeatter performed competent work and went on to count themselves among the nation's first professional wildlife managers. But it was Errington who ran with Stoddard's work on predator-prey relations to further challenge the mainstream understandings of both scientists and laypersons about predation.

Born in Bruce, South Dakota, in 1902, Paul Errington's background was not unlike Stoddard's. He spent his childhood in the outdoors, hunting, fishing, and developing an appreciation for nature in general. As a young man, he put himself through college with a trapping business, an occupation dependent on an intimate knowledge of animal behavior. Of Errington's ability in the field, Stoddard later reflected, "All too often scientists are conspicuously lacking in woodsmanship and are likely to consider it unimportant. Errington, who was an expert in the woods, was a striking exception to this generality."[18] Stoddard's judgment of potential wildlife students was rooted in his own experience as a museum field surveyor, when he might be out in the field for weeks at a time. Knowing one's way around the woods, fields, prairies, and wetlands, in other words, was indispensable.

Stoddard and Leopold helped Errington devise a study that was much more narrowly defined than Stoddard's. Errington's purpose was to measure winter quail mortality and to determine the influence of predators on quail populations on several well-defined plots of land. From the beginning, Errington pursued predation with a singular focus, and Stoddard noted on one inspection that "he is pushing this phase of his work with great energy and obtaining some valuable information." Indeed, Errington was so focused on his work that he did "not seem particularly open to suggestion" from his advisors in Madison. Stoddard went on to assure his contacts at the Biological Survey: "As he gives promise of developing into a very valuable man this trait of character is being overlooked as much as possible."[19] This was actually a "trait of character" Stoddard could appreciate; he was in the middle of his scrap over fire, which required ignoring one suggestion after another. Errington knew where he wanted to take the study, and neither Stoddard nor Leopold could get too upset about that kind of commitment.

While Errington got settled into his Wisconsin study, Stoddard was finishing off his quail manuscript. As will become clear, Errington was never quite satisfied with Stoddard's findings on predator-prey relations in *The Bobwhite Quail*, but it still served as a foundational text from which all future work, including his own, proceeded. In it, Stoddard crafted a complex yet elegant argument that placed predators within their ecological context. In making the case for wildlife management, one of his most significant steps was to place humans within a story typically reserved for animal interactions. Indeed,

Stoddard was explicit about substituting the "services" of hunters for those of the more rapacious predators of quail. Under natural conditions, where "cover and food supply are adequate," he wrote, "great reproductive powers usually enable the bobwhites to maintain themselves against their natural enemies. When man enters into the equation, however, and harvests from 5 to 25 per cent or more of the bobwhites' total increase it soon becomes evident that control of enemies is required to offset this unnatural drain." As shooting increased and agricultural conditions became less favorable for the birds, a quail enthusiast could "no longer put up his gun at the end of the shooting season and forget about the game until another season rolls around."[20] The close management of the environment was essential, and part of that management was the targeted, knowledgeable control of predators.

Stoddard's take on predator control, however, was far from that of the earlier control programs. It was firmly rooted in his background as a naturalist and ornithologist, yet his personal charge was to help reconcile the eradicationists and protectionists through science:

> When man is compelled to regulate other forms of life to his own advantage, he should do so in a reasonable, humane way, after first weighing all the available evidence, and when there is doubt, it is well to give the living creature the benefit of it. False propaganda, disregard of the results of scientific research, and over-zeal, coupled with a lack of humanity in controlling wild life, may be expected to 'backfire' and do more harm than good to the cause of field sports.[21]

This blend of science and morality was a driving force in the development of wildlife management, and the value of predator-prey study was not just more measured predator control. By following the connections between predator and prey, wildlife scientists could expose the interconnections found in nature and demonstrate that human action toward one species could have unintended consequences for others. Such interconnection had moral implications for human interaction with the environment, and the first step toward making them explicit was to build a base of knowledge.

A repudiation of predator control, however, was not the aim of wildlife scientists like Stoddard and Leopold. Stoddard, for instance, believed that all animals had intrinsic worth, but he also realized that humans had to engage nature and be a part of the equation. The first step toward that goal was to discover the workings of nature. In the quail study, for example, his two primary tasks in regard to predators were to identify which animals preyed most intensively on quail and to develop methods for their control. Throughout the course of the quail investigation, Stoddard examined and tallied the stomach contents of mammalian and avian species brought in by tenants from across the Red Hills; he observed predatory activity on captive birds; and, perhaps most importantly, he located a tremendous sampling of quail nests and made daily rounds to note any predatory activity.

After four years of intensive study, Stoddard came to several landmark conclusions that dispelled many myths about predators, and simultaneously raised as many questions. He was, however, reluctant to offer many theoretical generalizations. He recognized a "saturation point," or carrying capacity, that quail populations could reach depending on habitat availability, seasonal weather conditions, and predatory activity. But he did not formulate an explanatory apparatus to show how these natural mechanisms functioned. He also argued that hunters themselves should be included as predators, and that they should be controlled as such, thus making the regulation of both human and animal predators a necessary component of wildlife management. It was here, in the *control* of predators, that Stoddard made one of his most important marks. Just as he did with using fire to sustain the food supply, he preferred making the environment advantageous to quail survival instead of waging unscrupulous warfare on predatory species. The point of his research was to produce more quail, not to reduce predator populations. It was a distinctly ecological approach, but even under the best environmental conditions, he did not rule out the possibility of quail decimation due to certain predatory species.

The Bobwhite Quail systematically addressed each alleged quail predator. Stoddard first noted what most admirers of quail already knew—they were crafty in evading pursuers. They were "extremely alert birds, speedy of wing, and able to take advantage of every bit of cover the country affords."[22] Their very capacity to evade predators was what made hunting quail such good sport. Despite this common knowledge, most hunters still thought quail no match for the more aggressive predators. As an ornithologist, Stoddard had a special interest in setting the record straight on winged predators. Through stomach and pellet analysis, along with daily field observation, he determined that few birds of prey attacked quail on a regular basis. Of red-tailed hawks, which "have been so greatly reduced in numbers by farmers and hunters that only enough are left to add variety to the bird life of rural districts," he did "not recommend their destruction anywhere to aid in quail preservation, for the good they do more than offsets the harm." Likewise, the time given to killing red-shouldered hawks was "worse than merely wasted, for no instance of quail destruction on their part came to light."[23] Nor did owls present a threat: "One of the most discouraging sights seen by us . . . was a beautiful barn owl, hanging by one foot from a steel trap on a quail preserve where cotton rats were abundant. The services of this bird would have been highly valuable to the owner, but its life had been sacrificed by the indiscriminate pole trap, which has no place on the well-managed quail preserve."[24] The pole trap, an import from English hunting estates, was simply a steel jaw trap mounted on top of a bare pole, the type of perch favored by many birds of prey. In Stoddard's view, such an arbitrary control technique had no place in game management. Other birds of prey, including the sparrow hawk (American kestrel), broad-winged hawk, and marsh hawk (northern harrier), did not warrant any control, Stoddard insisted, their presence also being more often beneficial to quail than not.

There were, however, two problematic hawks. The Cooper's and the sharp-shinned, called "blue-darters" because of their bluish plumage and a tendency to quickly navigate the tight quarters of the woodlands, were a thorn in Stoddard's bird-loving side. The Cooper's hawk, in particular, was a "true bird-killing hawk" and was "largely responsible for the ill repute of the whole family."[25] Even with abundant cover—the most effective measure of environmental control for other winged predators—the blue-darters rarely missed their mark. Stoddard's tone turned unusually harsh when discussing these predators. He likened them to "the picturesque pirates of old . . . too violent and bloodthirsty to be willingly tolerated." In contrast to other hawks, the blue-darters feasted on the ground-dwelling bobwhite quail: "These destructive hawks probably harvest a crop of quail in the aggregate during the course of their 365-day open season comparable with that taken by sportsmen in their much shorter time afield."[26]

The problem, in other words, was not the blue-darters per se; it was that quail populations could not withstand the voracious appetites of both human hunters and blue-darters. Under "natural" conditions the blue-darters performed their duty admirably as a check on quail populations, but the unavoidable reality was that hunters had priority on the Red Hills preserves. Such reality made Stoddard unwilling to embrace categorically biocentric imperatives in *The Bobwhite Quail*. But he was still reluctant to support a campaign against blue-darters—he saw little inclination in the sporting public to distinguish them from other hawk species. Nor did he think a targeted campaign would do much good—they were simply too elusive. Instead, he hoped to present a more expansive view of predator control in terms of the aggregate environment, rather than specific predator-prey relationships.

When observed in isolation from other predatory species, avian predators presented a fairly straightforward picture, but when mammalian and reptilian predators entered the mix, all sorts of complexities arose. Cotton rats, skunks, opossums, foxes, bobcats, house cats, black snakes, coachwhip snakes, diamondback rattlesnakes, and more all fed on bobwhite quail at some point or another. After conducting hundreds of nest studies, Stoddard wondered how a young quail "could ever run the often bewildering gauntlet of dangers and reach maturity."[27] Over half of all nesting attempts were unsuccessful because of natural enemies or heavy rains, and once young quail began feeding on their own they had to evade the blue-darters. It was a tight spot, but quail had some important adaptive strategies that helped them carry on. After they abandoned a nest for whatever reason, the seasonally monogamous pair continued to nest again and again, and Stoddard estimated that the great majority of pairs eventually hatched a brood. He also estimated that twelve chicks hatched in the average brood, making for a substantial increase from the initial pair.[28] In a well-populated territory like the Red Hills, only one pair had to survive the year to reoccupy the breeding range. Predators, in other words, performed the important service of preventing overpopulation.

The dilemma for wildlife management, however, was how to add hunters to the predatory mix. Stoddard feared that "conditions in some areas are undoubtedly so evenly balanced between reproduction and average natural mortality that the shooting of any bobwhites throws the balance against the species and causes a decline."[29] Predator control was a necessity, but by control he did not mean the killing of individual species. Instead, he meant environmental control. Part of the equation was to create environmental conditions favorable to quail, as was done with fire-maintained open forests and tenant agriculture. With plenty of food and plenty of places to hide, quail had a fighting chance against species that preyed on juveniles and adults. Those species that preyed on quail eggs, though, presented a greater challenge. Cotton rats were a particular problem. Zoologists commonly thought of them as a "buffer" species, or one that diverted the attention of promiscuous predators away from more desirable species such as quail. As he spent more time observing the Red Hills preserves, however, Stoddard noted he increasingly had "no question in my mind that on southeastern quail lands the 'buffer' theory is a delusion and a snare."[30] Besides feeding on quail eggs themselves, Stoddard argued, their presence in large numbers attracted even more predators that fed on quail incidentally. The cotton rat, then, was an important link; diminish its population, and that of other predatory species would fall concurrently. Though Stoddard experimented with poisons, "the best means of preventing them from becoming numerous on quail preserves," he insisted, was "to burn carefully."[31] They thrived in heavy rough, but not in frequently burned woodlands. Again, the use of fire was the most important step toward land management in the southern coastal plain. If he could make the environment as inhospitable as possible for some predatory species, like the cotton rat, then he could leave alone other species that were more difficult to control, like the blue-darter hawks. Even still, the primary goal was to make the environment hospitable for quail, which was bound to attract their enemies.

The complex human ecology of the Red Hills preserves often aided in predator control efforts. One way to hedge the bet in favor of quail was the trapping and night hunting of small mammals, a generations-old activity that Stoddard considered in perfect keeping with his idea of environmental control. The fur trade in raccoons, skunks, and foxes maintained an economic foothold in the early twentieth century, and many small mammals, like opossums, were an important food source for tenant families. Indeed, Stoddard thought the widespread hunting of these animals an important reason the South became such a desirable location for quail hunters in the first place, and it was another important cultural dimension to stable quail populations.

Landowner efforts to ward off trespassers, however, complicated the night hunting of small mammals, thus threatening to undermine this effective form of predator control. As they had when originally developing the preserves, landowners continued to define and redefine their property boundaries with fences and posted signs in an effort to close the open range. Stoddard became concerned such steps would harm quail populations. He worried that "posting and patrolling the property usually involved prevention of commercial trapping of fur-bearers," as well as curtailed the "old southern sport of 'possum hunting,' which had held the opossums, skunks, and raccoons in check since pioneer days." Without these activities, "many of these quail preserves were in effect 'vermin sanctuaries.'"[32] Stoddard's use of the word "vermin" suggests the undervalued status of small mammals in wildlife circles, but the overall point was to prevent population irruptions of any one species, whether the esteemed quail or the lowly skunk. More importantly, in positioning some forms of human predation as critical pieces of the ecological puzzle, Stoddard crafted a model of wildlife science that included, indeed mandated, human intervention in ecological processes.

The management of such human interventions took some coordination between property owners. Though it is unlikely many preserves stopped local tenants from "possum hunting," it is clear that the lines drawn by posted property boundaries complicated such activity. The night hunting of small mammals, as practiced with dogs, required a great deal of territory and pursuers frequently criss-crossed property lines with little compunction. The Red Hills preserves, however, were in a unique position to continue this open-range tradition regardless of posted property. As discussed in chapter 1, they composed one whole unit that encompassed a great deal of land, and since "many of the fur-bearing enemies of bobwhites are animals that wander extensively, cooperation from surrounding landowners is essential to their effective control. For this and many other reasons, groups of quail preserves have a great advantage over those that are isolated."[33] Preserve owners could easily coordinate such control measures, thus allowing for clear passage across property lines. Perhaps just as important, and more than a mere incidental benefit of small mammal hunting, "the pursuit and utilization of fur-bearing enemies of quail may be made to furnish both pleasure and profit to preserve employees."[34] The economic and recreational activities of tenants, in other words, were crucial to maintaining the preferred recreational conditions of the preserve owners. Quail hunting, then, did not always dispossess marginal hunters; just as often it profited from their activities.

The public greeted Stoddard's results with a considerable amount of interest. The *New York Times* ran several stories on the study and paid particular attention to Stoddard's findings on predators. One story quoted the curator of birds at the American Museum of Natural History, Waldron De Witt Miller, saying that Stoddard's study "should satisfy the most exacting that the food habits of hawks and owls as a class are such as to make them of the greatest benefit to man."[35] In another piece, the *Times*'s "Rod and Gun" columnist, Lincoln Werden, noted Stoddard "pointed out that much harm is done unintentionally by those who unknowingly destroy birds which are not actually destructive," and goes on to quote Stoddard at length.[36] The *Times*'s review of *The Bobwhite Quail* even went so far as to call it "the last word on the bob-white quail."[37] Hyperbolic, perhaps, but Stoddard's work was clearly of interest not only to scientists, but the sporting public as well.

By the time Stoddard finished the quail study he had an overwhelming body of evidence on the habits of predatory species, and he had worked out complex methods for environmental control. But a reluctance to generalize prevented him from offering a more sweeping theory of predator-prey relations. Two years after *The Bobwhite Quail*'s publication, Aldo Leopold's *Game Management* was an even more far-reaching work that made a strong, concise argument for more rigorous management of wildlife habitat. As game habitat dwindled and public demand for gun sports increased, he built on Stoddard's study to stress the need to understand game management as one of the "land-cropping arts," much like forestry and agriculture.[38] Leopold's approach to predator control was as practical as Stoddard's. In attempting to mediate the conflicting interests of four groups—agriculturalists, game managers and sportsmen, students of natural history, and the fur industry—Leopold pointed to Stoddard's work as proof that the "actual measurement of losses from predators is thoroughly feasible."[39] He encouraged sportsmen, in particular, to use science as their guide in identifying and culling predators, and echoed Stoddard in his contention that "cover and food is a better protective measure against some types than the killing of the predator."[40] Stoddard thought *Game Management* went further to expand the field than any work before it, and upon reading the manuscript, wrote, "It should do a great deal to aid those working with game to *think straight* and I particularly hope that every hunter in America reads it, for you point out in a singularly understandable way that there is something more to increasing and maintaining game than exterminating 'vermin' and closing shooting seasons."[41] Leopold's text, coupled with *The Bobwhite Quail*, alerted the public not only to the nation's diminishing wildlife habitat but also to the notion that the restoration of habitat was a more felicitous method of game protection.

While the scientific community and sporting public reacted to Stoddard's and Leopold's texts, Paul Errington began publishing a series of articles on his quail predator studies that pushed the field even further toward an ecological perspective of conservation. Building on Waldo McAtee's assertion that "predation tends to be in proportion to population," Errington argued that when prey populations exceeded the carrying capacity of favorable habitat, predatory activity surged proportionally to prevent overcrowding; when the prey dipped below carrying capacity, predation slowed down as well, thus allowing populations to regain numbers. "In other words," he wrote, "if a quail population fits well into an environment it suffers light or negligible loss from predation."[42] The key to maintaining a game species like quail, then, was the control of habitat, not the control of predators. Under favorable environmental conditions, predation was simply an incidental occurrence rather than the determinant of quail numbers. These results had serious implications for the conservation community; Errington's message was to leave predators alone completely. To those who continued to target predatory species, he urged: "The obvious trend of modern ecological data is toward the conclusion that predation does not play nearly the part in determining population levels of wild species as was thought a comparatively few years ago."[43]

Stoddard was wholly in sympathy with Errington's sentiment but felt his interpretations outpaced the actual data. In a long and detailed correspondence they shared with McAtee and Leopold, Stoddard and Errington hashed out their disagreements over Errington's dissertation and related publications, eventually reaching a compromise with a coauthored article in 1938. One of Errington's primary revisions to *The Bobwhite Quail* was to abolish the remnant value judgments that Stoddard used in assessing predators. No longer were Cooper's hawks bad, red-tailed hawks good, and coachwhip snakes a middling species somewhere in between; the standards of human moral judgment, in Errington's estimation, did not apply to the world of predators. He insisted that if one predator didn't reduce the quail surplus, another would. There was no point, in other words, to the targeted control of any one species. As he explained it to McAtee, "Conspicuous mortality of many kinds may not be of any actual significance except to inflame the resentment of our own jealous species. . . . It seems to make scant measurable difference what native predators are in the environment nor how abundant they are as long as the environment from the standpoint of the quail is strong otherwise."[44]

This was a contention that Stoddard "utterly fail[ed] to grasp or agree with."[45] His primary job in the Red Hills was to ensure shootable numbers of quail year after year, and his experience told him that an influx of predators like the Cooper's hawk could diminish quail populations with much greater efficiency than furred or scaled predators. When drought damages the ground cover over a heavily stocked territory, Stoddard noted as an example, "Coopers Hawks drifting through on their migrations discover the extreme availability of the birds, harry them through the Winter and severely reduce the populations. In this case one predator, rather than the whole environment seems to be mainly responsible."[46] Once again, Stoddard's work was closely aligned with the practicalities of wildlife management. His management program began with a general approach and adapted throughout the seasons to particular environmental changes. In comparing Errington's science with his own management, Stoddard continued:

> In managing quail preserves in this part of the world we depend first on balancing food supply and cover, and increasing one or the other or both where practicable, and only control predators where, when and if necessary. But to ignore the predator factor would be very unwise in many cases. One constantly has to endeavor to maintain a balance most favorable to the quail, without doing anything that might be against the welfare of the country as a whole. For this reason I contend that constant study of *detail* is necessary to the game manager, as well as a general grasp of the principals [*sic*] involved . . . general principals [*sic*] only become evident with the study of vast quantities of data. I have insufficient data to prove anything though they may indicate trends.[47]

Local environmental conditions dictated Stoddard's approach to wildlife conservation, as did the need to maintain quail abundance. Both Cooper's hawks and humans were predators in Stoddard's estimation, and they both warranted control if they threatened that goal.

Upon receiving Stoddard's criticism, Errington found himself "sunk by a consciousness of the utter futility of trying to discuss complex things by correspondence."[48] Stoddard had been an important influence, and Errington was not interested in discrediting the important work of a friend. He was persistent, nonetheless, and did his best to explain his interpretation of the data. Errington first wanted to dispel any notion that he overlooked some evidence in an attempt to stop predator control outright. He claimed no special reverence for predatory species, and he hesitated to overturn common biological principles that held predators to be a major factor in determining prey populations. At the same time, his data was so insistent "that I don't see how they can be disregarded unless the data themselves are all wet—which I surely don't think to be the case, as they were gathered and published year by year without the remotest idea on my part as to what they would ultimately signify when considered collectively."[49] His research was thorough and meticulous, and the evidence clearly demonstrated quail predators to be little more than a nuisance.

This was in Wisconsin, however, on the northern edge of the bobwhite quail's range where environmental conditions were good and hunting pressure very light. Human predators were not a part of his ecological model at all. Stoddard, on the other hand, worked in landscapes that were maintained with the express purpose of hunting; he simply could not fathom the utility in postulating a general theory of predator-prey relations without including human predation, especially of a game species. He replied to Errington that "the man with the gun can reduce quail in a favorable environment more quickly and efficiently than any other predator, but to permit him to reduce them to the greatest degree without curtailing future privileges in this line is the aim of nine tenths of the research and management to date. It is about all I am doing with quail, much as I love them."[50] The purpose of wildlife management, Stoddard thought, was to develop a means for wildlife to cope with human pressures. And while hunting seasons and bag limits were practical controls on the human predator, the eradication of hunters was not. Wildlife management, then, did little good in the South without some account of human ecology. In Errington's Wisconsin, on the other hand, intense hunting pressure was not part of the ecological reality, so he saw little need to include people in his study model.

Stoddard and Errington let the dust settle for a few years until they could sit down together and attempt to reconcile their differences. As time passed, Errington became increasingly aware that many readers sensed serious differences between his interpretations and Stoddard's. The conference circuit must have been buzzing with scuttlebutt about these new views of predator control and their architects, and Errington was ready to clear the air. He expressed his concern to Waldo McAtee, writing, "There is a tendency, especially on the part of younger men, to assume that when two investigators report conflicting findings that one or the other has to be wrong."[51] Errington did not think this had to be the case, but he and Stoddard did not try to resolve anything until they could talk it over in person. They finally met in the summer of 1937, most likely at Errington's home in Ames, Iowa, where he had accepted a position at Iowa State University.[52] After a long day of discussion, they recognized their disparate findings to be a matter of place, climate, and management goals, rather than of differing principles. Errington was relieved, and he immediately started writing an article based on their meeting. He hoped that a joint statement on predator-prey relations might "show certain people who read more than is written into my writings that we are in some sort of accord in matters of consequence," as well as "possibly clear up a vast amount of misunderstanding already in existence."[53]

In the paper, published under both of their names, they acknowledged the apparent differences in their previous work, and proposed some ideas about why predators influenced quail populations in the South more than in the Midwest. They speculated that a higher density and variety of common prey species like mice, ground squirrels, rabbits, and large insects in Wisconsin caused less predation on quail since predators had so many food sources.

The buffer theory, in other words, held its own in the upper Midwest. But in the Southeast, cotton rats were the primary buffer species, and their local populations could fluctuate so dramatically that a sharp drop in numbers could send their predators looking for the next available food item, which "may seriously affect bobwhite populations."[54] They both freely admitted that "the causes of these regional differences in predation and population phenomena are still too obscured by unknowns to permit full explanation," but the point of the article was to verify regional differences, and to present a unified front against indiscriminant predator control.[55]

Oddly, though, they did not recognize some fundamental differences between their respective studies, at least in this article. Stoddard never hesitated to intervene in ecological processes when his fieldwork suggested it was needed. He burned frequently, directed agricultural practice, planted vegetation for feed, and, yes, shot or trapped predators when he thought it necessary. He was an active participant with his research subjects. Errington, on the other hand, refused to intervene in ecological processes. Though his research landscape was shaped by human activity, his goal was to observe and report the predator-prey dynamic as it played out on the wild edges of agriculture, not to help shape it. In some ways, their research design reflected their circumstances. Hunters were an important predatory influence on the Red Hills preserves, so Stoddard sought to devise ways for hunters to compensate for their influence in responsible ways. And since human predators were virtually absent in Errington's landscape, he had no need to experiment with habitat and predator control. Their social and ecological surroundings, then, helped to shape their research design, as well as their conclusions.

While Stoddard and Errington were able to resolve their problems by pointing to regional peculiarities, their differing approaches to field research reflected a broader identity crisis within the developing profession of wildlife management. In the decade or so after Leopold and Stoddard laid the foundation for the field, no one struggled much with defining it or with structuring its professional parameters. The creation of a professional organization devoted to the subject in the late 1930s, then, was a crucial moment for wildlife management. As more universities adopted the model set up by the SAAMI fellowships, and New Deal programs like the Soil Conservation Service and the Resettlement Administration created a multitude of outlets through which to practice wildlife management, it began to take its modern form as a feasible profession. But there was still little consensus on the driving purpose of the field. Most everyone recognized its practical origins and agreed that it should serve as a guide to wildlife policy, both public and private; but as it became further entangled with the various subdisciplines of academic biology, a schism between the managers and the scientists developed that closely mirrored Stoddard and Errington's differing approaches to predator-prey research.

Creating a new professional group meant the creation of a new professional identity, and the consolidation of its core membership was the first step. As a hybrid field, wildlife management pulled scientists from a number of biological specialties, as well as attracted the attention of laypersons from the many policy-oriented wildlife groups that proliferated in the early to mid-twentieth century. In an effort to distance themselves from the laity, as well as to gain much-needed academic respectability, a group of thirty-four met at the 1936 North American Wildlife Conference in Washington, D.C., to discuss building an organization for specialists working explicitly in wildlife science and management. According to McAtee, "This action was prompted by the conviction that there is need in this rapidly growing profession for an agency to define and maintain standards for the work and workers, to affiliate for the common good all subscribing to those standards, and to consider the feasibility of establishing a periodical devoted to wildlife management."[56] The response was enthusiastic, and the recognized leaders of the field immediately started a flurry of correspondence to work out a structural organization.

The first task—not as mundane as it may seem—was to name the group. The name would communicate the constituency of the organization as well as the purpose and agenda of the discipline. Most agreed that the temporary name, the Society of Wildlife Specialists, was uninspired. Nor would the Society of Wildlife Managers be sufficient—the organization included specialists in pathology, botany, zoology, forestry, and more. As the core organizers considered an apt moniker, one of Leopold's students at the University of Wisconsin, Leonard Wing, thought up a professional title for wildlife experts that caught McAtee's eye. In an essay in the American Forestry Association's *American Forests*, in which Wing urged the forestry profession to "naturalize the forest for wildlife," he passively observed that "the conservation biologist is a new entry into the field of conservation."[57] This, thought McAtee, was what they were: conservation biologists! He quickly wrote Leopold and Stoddard, saying, "Has not Wing given us just the term we need, and which it is surprising no one thought of before, namely, Conservation Biologists?"[58]

They agreed completely. It best reflected their use of scientific research in service of conservation, and their identity as professionals addressing questions about wildlife conservation through the methods of sound science. It also created some distance from hunters, thus widening their perspective to become a group not simply concerned with wildlife for an anthropocentric purpose. For McAtee, the name "seemed to me to complete the picture," and covered "everything we have in mind in connection with our Society, is dignified, and places the emphasis as it should be, upon conservation."[59] While this discussion of "conservation biologists" reveals how this core group began to define themselves, the name, for reasons lost in the available record, fell by the wayside.[60] At the first official meeting of the new organization in 1937, they adopted instead the rather ordinary "Wildlife Society."[61]

The overlap of wildlife management with a variety of other disciplines was bound to create some tension. Some questioned, in particular, whether wildlife management needed its own journal. One proposal worth discussion came from Wallace Grange, a former Biological Survey employee who worked closely with Stoddard and Leopold on the SAAMI project, and later established a commercial game propagation farm in Wisconsin. Grange recommended the Wildlife Society publish a technical section in *Game Breeder and Sportsman*, a magazine, just as its name suggests, devoted to those who raised game animals to release for sport. In making his proposal, Grange exposed a scale of artificiality within wildlife management he thought should be abolished. What were wildlife managers, he suggested, but animal breeders? He pushed Leopold on the subject, writing, "In my opinion game breeders are game managers, and game breeding is management. . . . As a matter of fact, game breeders are technically the *only* game managers. Others are *land* managers, *man* managers, *gun* managers, but not actually game managers."[62] Grange had a point—wildlife managers created environments for animals to breed much like a propagation plant. The argument, however, challenged the very foundations of wildlife management, and neither Leopold, Stoddard, nor McAtee gave it much of a hearing. Leopold's views on the subject were well known by this time—not only did wild game make for better sport, he had argued, it was an inherently more democratic approach to management.[63] McAtee dismissed the proposal outright, chiding Grange that this was an organization for "those genuinely interested in wildlife management and wildlife research with conservation, not exploitation, of wildlife as a background."[64] Stoddard was a bit more reflective, writing that game breeders had their place in the scheme of American wildlife management, but that the Wildlife Society had a more expansive view of animals and their environments; it was to be a home to "a rather large group who are in the business of trying to increase or maintain wildlife by manipulation of environment, as distinct from attaining the same objectives by means of artificial production." A further contrast was that "the artificial breeder of game is primarily interested in game. The organization we are building would include the agencies who are managing wildlife, as distinct from game only."[65] The key term here is "wild." Breeders, according to wildlife managers like Stoddard, were propagators who had little interest in maintaining and restoring wild habitats. They raised game animals in "plants" and released them for sport—the antithesis of wildlife management.

On the other end of the spectrum was a proposal to affiliate with the Ecological Society of America (ESA), the only proposition to gain legs among the founding members. Established in 1915 as an organization "for the purpose of giving increased unity to the study of organisms in relation to environment," the ESA spoke a language similar to that of wildlife specialists.[66] Indeed, a marriage of the two groups seemed like a natural move. Like ecology, wildlife management was a field that viewed environments as a series of interdependent parts, rather than simply a group of individual plant and animal species. Wildlife specialists paid close attention to vegetative succession and the interrelations of animal populations, the very marrow of ecology. Walter Taylor, a Biological Survey employee and member of the ESA executive committee, first drafted the idea, and it received strong support from ESA's president, W. S. Cooper, a biologist at the University of Minnesota. Leopold, Errington, and Ralph King, the first president of the Wildlife Society, met with Cooper in November 1936 to discuss the details of the merger. Going into the meeting, Errington said he was "on the fence, and I think that Leopold's views were dubious. After an afternoon's joint session, we were all very much in favor of affiliation." Errington felt certain that "if such affiliation came to pass I don't see how the [Wildlife Society] would be dominated by anybody, that it would lose its identity or that it would suffer any apparent disadvantages. . . . I think greater scientific standing would be gained for our group. I am also convinced that affiliation would further the attainment of what are in actuality our joint objectives—conservation and science."[67] In fact, the conservation goals of the ESA were not at all clear. They were in the middle of a long controversy of their own that questioned whether or not scientists should be engaged in environmental advocacy, a conflict that caused the animal ecologist and founding president of the ESA

Victor Shelford to eventually depart and form his own conservation organization, which became the highly influential Nature Conservancy.[68]

It was just such a division that worried Stoddard and McAtee about the proposed merger, though they considered the issue in somewhat different terms. McAtee worried that the open nature of the ESA allowed the infiltration of political appointees, something he thought would forever ruin conservation-based science. Stoddard also favored a separate organization with its own journal, though he understood both sides of the issue. On the one hand, he recognized an inherent connection between the two disciplines: "We all fully appreciate that our work may be 'applied ecology.'"[69] On the other hand, he thought the research of ecologists had "little real application to wildlife management." He hoped, instead, to see research "having an important bearing on *management*" occupy its own organizational and literary space.[70] He feared, most of all, that too much attention given to ecological theory would draw wildlife managers away from the environmental contingencies they dealt with on a daily basis. He was not so concerned with losing the rights of advocacy, per se; rather, he was anxious about the continued ability to shape environments according to desired goals.

As the merger proposal made its rounds through correspondence, Stoddard's rhetoric grew more forceful, and he related the plan to a trend he already sensed in wildlife management:

> I have a very real fear that having long gone to one extreme in so-called "vermin control," we are now headed for the opposite extreme, following the general American tendency of never following a moderate or middle ground. My whole field experience indicates that certain theories now being followed relating to predation are unsound, and that time will *prove* them unsound from any practical standpoint. If the wildlife management movement does not recognize this and have a care, our organization will soon be regarded by a much more powerful group as a bunch of long haired theorists, and the result will be a wider rift than ever in the conservation field.[71]

These were interesting words from someone whose work would later be embraced by "a bunch of long haired theorists," but his sentiment is revealing of the moderate stance he hoped all wildlife managers would take when navigating theory and practice. More to the point, he thought theory was too ethereal and static to guide a manager in the field. Stoddard felt strongly "that the ecological picture can never be completed by man, and that it is changing all the time"; reliance on a theory developed under one set of conditions, then, might hamstring someone managing land under a different set of conditions.[72] He kept close tabs on the scientific literature, but still preferred learning by practice.

Despite their disapproval of joining up with the ESA, Stoddard and McAtee receded to the background in this particular debate. They did not attend the 1937 meeting of the Wildlife Society in St. Louis, where the membership voted on affiliation. Stoddard, by now approaching fifty years old, wanted the Wildlife Society to be "a youngster outfit," and would support whatever the majority decided, while McAtee simply wanted to maintain good relations with Paul Errington. They did, however, have representation at the meeting. Perhaps one of the most vocal "youngsters" in opposition to affiliation was one Stoddard helped to train, Edwin V. Komarek. A former student of the animal ecologist W. C. Allee at the University of Chicago, Komarek first visited Stoddard in 1933 while on a small mammal collecting trip for the Chicago Academy of Sciences. Komarek's skill and enthusiasm were impressive and Stoddard soon hired him as an assistant.

By the time of the St. Louis meeting, Komarek had fully embraced Stoddard's approach to wildlife management, and he became Stoddard's proxy on the question of affiliation with the ESA. Komarek, forever loyal to Allee, his early mentor, realized "that the basis of wildlife management is ecology," but wondered rhetorically to Walter Taylor "why the wildlife field has grown outside of the Ecological Society. Perhaps the answer is that ecologists did not want to see the *practical application* of ecology." An overly general claim, to be sure, but Komarek's point was to avoid alienating the constituency of wildlife management. He continued, writing "that the basic principal for the existence of wildlife research is better farming, or management, of wildlife not a quandary of theories and scientific terms. I dare say that few sportsmen, and that's the group we are all working for, know even the word ecology. Let's not get them scared of us like farmers are of agricultural scientists."[73] It was important for both Komarek and Stoddard that wildlife research be easily translated for the public, and while personally and professionally interested in ecology, they rejected what they saw as the insularity of academia. Heading into the St. Louis meeting, this argument gained some momentum, and the decision did not create the schism that Stoddard and McAtee expected. Some participants publicly accused the ESA of "intellectual snobbery," and the Wildlife Society membership voted against affiliation with only two dissenting votes.[74] They also created the Journal of Wildlife Management, and nominated McAtee as its editor.

With this meeting behind them, the Wildlife Society set out to create an intellectually rigorous, well-defined process for becoming a professional wildlife manager—that is, they became an institutional agent of professionalization. The committee to do so consisted of Stoddard, Errington, Walter Taylor, and fisheries biologist Carl Hubbs, but it was Aldo Leopold, who "for years has been thoroughly fed up with the red tape" of other professional societies, who took this task as his own.[75] Leopold worked on the proposal throughout 1938, outlining the basic training needed for wildlife work. Wildlife education, according to Leopold, had two ambitious goals: "1) to teach a few men to make a living by managing wildlife; 2) to teach the whole body of citizens to appreciate and understand wildlife." In developing the steps to fulfill those goals, Leopold chose to focus on the general characteristics of a wildlife professional—"what the student is, what he knows, what he can do, and how he thinks"—rather than the specifics of university training.[76] As various drafts of the proposal made their rounds, this general approach rubbed some the wrong way. According to one critic, it gave neither the university professor nor the prospective student proper guidance in constructing a curriculum for wildlife management.[77]

In response to such criticism Leopold leaned heavily on Stoddard, whose views on professional training were well known to the Wildlife Society membership. Using his own background as a touchstone, Stoddard had already argued vehemently that practicing wildlife managers be awarded active membership in the society, regardless of educational qualifications. He carried this further in assessing proper educational guidelines in the academic sphere, telling Leopold that "I am a firm believer that the function of an institution of higher learning (a detached viewpoint, as I have never personally attended such an institution as you well know) is to teach the student to *think*, and to help him find out how to use the world's accumulated knowledge to the best advantage."[78] The problem with university training, for Stoddard, was not the coursework per se; it was that courses fell under the umbrella of the *institution*. He analogized the university to the church, writing,

> A man may be *aided* in becoming a good Christian by listening to the preacher, but after all he has the same sources of information (no more, no less) as the preacher, and can get the information himself if he has the proper mental powers and enthusiasm, without ever seeing the inside of a church. . . . In a profession *where a student and his teachers must learn together* (is there any other established profession with only one text book?) courses in my opinion should come second to field studies in the open.[79]

It was in the field, according to Stoddard, where the wildlife student learned. Both Leopold and Stoddard knew burdensome institutional obligations could threaten to overwhelm the original purpose of study. They also knew professional momentum might carry wildlife management far from its roots in the field. Leopold took Stoddard's words to heart. In answering the critics of the professional standards report, Leopold quoted "for the edification of the committee a sentence from Stoddard's letter: 'Are we not making an undue showing of our immaturity as a profession by all this talk of courses, universities and so forth, as though they were the aim of wildlife management, rather than a desirable transitory period in the life of a wildlife manager?'"[80]

As it turned out, Leopold's finished statement on professional standards was not a detailed guide to constructing a wildlife curriculum, but it did make clear that the path to becoming a wildlife manager passed directly through the university.[81] A professional wildlife manager now needed four years of undergraduate study and at least one year of graduate training. No longer could an apprentice taxidermist and self-taught ornithologist like Stoddard rise through the professional ranks. In fact, the great irony of Stoddard's career is found in his very success. Despite preaching the importance of practical training and "woodsmanship" his entire life, he was partly responsible for the establishment of wildlife management as a formal discipline and profession, thus making his sort of informal training obsolete as a path into the profession. By helping to create the profession, he also helped make it more difficult for people such as himself to reach the professional level.

The nature of the subject, however, ensured that wildlife management remained a field-based science, just as the complexities of the nature–culture interface ensured questions like those of the population dynamics of predators and prey remained difficult to answer. Stoddard, Errington, Leopold, and McAtee helped to construct a scientific framework through which the conservation-minded public could view predators ecologically, thus lending support to anti-eradication efforts, and eventually to the campaign for threatened and endangered species. They were helping to create a biologically centered view of the world. But by the early 1930s, most wildlife biologists were focused on protecting wild nature within human systems of production; and as they encountered specific environments and environmental problems, they increasingly found that there was no normative American environment, only a series of choices. They largely based their choices on a new ethical relationship with nature, but nature sometimes got in the way. And the specific questions of on-the-ground land management did not always comport with the generalist impulse of scientific inquiry. For a locally based land manager like Herbert Stoddard, for instance, nature itself was dynamic, and his management decisions came in response to climatic and land cover fluctuations (of both the human and nonhuman variety) more often than scientific or ethical paradigm shifts. Paul Errington, on the other hand, sought to construct an orderly model for animal population dynamics, one that could be applied universally. The organizational schism in the creation of the Wildlife Society, in turn, largely mimicked these disagreements over theory and practice. When theory threatened to limit practice, as well as define the profession, Stoddard drew on his own background in an attempt to scale back a professional agenda he considered overly ambitious. One way to do that was to engage the types of land-use patterns that dominated the southern and American countryside—those of agriculture. If wildlife management was to have any broad success in application, if its lessons were to reach beyond the academy and the government, it would need to influence agricultural landscapes.

CHAPTER 5

Wild Land in Cultivated Landscapes

As Herbert Stoddard put the final touches on *The Bobwhite Quail* in early 1930 from his temporary home in Washington, D.C., several of his closest acquaintances busily concocted plans for his future. He already acted as the Biological Survey's representative for the Sporting Arms and Ammunition Manufacturing Institute's game fellowship program, the first real attempt to insert wildlife research and management into the nation's universities. That program's leader, Aldo Leopold, considered Stoddard the nation's premier mind in wildlife research, and he had plans for Stoddard that included increased research activity from within the Biological Survey, a PhD degree, and eventually a university appointment as one of the nation's early professors of wildlife management. Waldo McAtee had also charted a course for Stoddard's immediate future. Stoddard would remain with the Biological Survey, replicating the Quail Investigation in other locations, testing many of its findings and tailoring management practices to suit the environmental and human peculiarities of place. With the help of Leopold and McAtee, Stoddard would become fully engaged with public-minded conservation, no longer limited by the scrutiny and control of the Quail Investigation's backers.[1]

His primary backer, however, had different plans for Stoddard. Lewis Thompson felt the work in the Red Hills had just begun and that land management on the local preserves would suffer tremendously in Stoddard's absence. Stoddard had expressed an interest in maintaining ties to the region, but he had given little thought to staying on full time. The Biological Survey was committed to devoting more resources to wildlife management, and Stoddard wanted in. But that was before Thompson presented him with a rather remarkable offer: the outright gift of Sherwood Plantation if only

Stoddard and his family would live there, an offer "very difficult for a poor man of my inclinations to ignore."[2]

In two short years, Stoddard and Leopold had developed a deep affection for each other, and upon receipt of Thompson's offer, Stoddard quickly wrote Leopold, seeking his counsel. Leopold's response was supportive, if tempered by a concern for the insular nature of the Red Hills quail preserve set. As for Stoddard's continued role in SAAMI, he hoped that "if Colonel Thompson will now bear in mind your value to the country, as well as to him, I am sure some mutually workable plan can be set up." Stoddard would maintain a title of collaborator with the Biological Survey, training his successor, inspecting the SAAMI research projects, and supervising the Survey's duplication of the Quail Investigation in other locations. Such a setup would bring "to a head the need of training recruits in the U.S.B.S., and the need of appropriations to do it on." Leopold also suggested that Stoddard "'hang out your shingle' as a consulting game manager," thus "demonstrating by actual example that a game manager is a practicable profession."

Despite his optimism, though, Leopold's tone hardened about the possible outcome of Stoddard's move into private consultation. Stoddard's knowledge and ability, he thought, could be more fully utilized in the public sphere. Stoddard's ability to navigate between public and private conservation was "all premised on the assumption that Col. Thompson does not need all of your time. To claim all of your time would, in my opinion, be a mistake on his part, and to give it, a mistake on yours. I make bold to say that his accomplishment in fathering the Georgia Investigation (as a public move) will be remembered long after his success in business will be forgotten. To now let it relapse into a mere private enterprise would be too bad. It has only started. His cue is to continue it, and make you available to spread it to 48 states."[3]

Stoddard shared Leopold's concern for the nation's landed resources, and felt the Red Hills would be as good a base as any for participating in a national discussion. Agricultural landscapes were of particular concern to wildlife managers like Stoddard and Leopold. Indeed, they were as interested in enhancing the environmental conditions of agricultural landscapes as they were in preserving or maintaining wild lands. It was no historical coincidence that their attention turned to the cultivated environment when it did. During and after the years of the quail study, American agriculture experienced dramatic changes. Across the country, new technologies and expertise combined with new legislation and credit arrangements to create what historian Deborah Fitzgerald calls the "industrial ideal in American agriculture." Small farms began to consolidate on an industrial business model, mechanization allowed for a more efficient use of manpower, and experts adopted a scientific rationale for efficient, modern, and clean systems of agriculture, all of which resembled the productive model of the industrial factory.[4] The environmental consequences of this transformation were manifold. Government technicians and individual farmers set out to drain swamps, divert streams, and clear forests; fields grew to unprecedented sizes; new chemical fertilizers and pesticides found their way into soils, water supplies, and human bodies; and a mass of people moved on to other pursuits. After World War I, American agriculture entered the modern age with little time for the mule-driven plow or free-ranging sow. As Stoddard wrote in *The Bobwhite Quail*, "This 'clean-up' policy has in many cases been fostered by some agricultural leaders in their zeal to assist the production of maximum crops, the potential game production value of the land being completely overlooked."[5]

The South adopted these changes later than the rest of the nation. Credit arrangements between planters, merchants, tenants, and sharecroppers provided little incentive for the capital investiture for mechanization, and the South did not fully embrace agricultural industrialization until after World War II. But the process was well underway in the 1920s and 1930s, and if Stoddard was privy to its early entreaties, he also recognized that he was in a position to mount a challenge. Throughout the region, the best capitalized planters purchased tractors and began to consolidate small farms into larger tracts, allowing them to do a more thorough job of clearing and cultivating fields with less labor, as well as push tenants and other small farmers off the land. The threat of the boll weevil kept farmers in a constant state of anxiety, and encouraged a cleaner, more manicured farmscape. Boll weevils wintered in the protective brush at field edges, and many agricultural experts argued that the elimination of such habitat was crucial to controlling the pest. Further still, the Depression and subsequent New Deal accelerated change in the South. Small landowners defaulted on bank loans and lost their farms while locally directed New Deal subsidies filtered to the largest planters, providing them a base for expansion and further depriving a tenant population that was nearing complete dispossession. Indeed, sharecropping and tenancy were on the wane, soon to be replaced by mechanization, with the landowner becoming more directly involved in daily farm operations through the direction of a handful of wage laborers. Farmers abandoned many of their less productive fields, allowing them to move through successional stages from broomsedge to old field pine and eventually to a dense tangle of hardwood and pine. Meanwhile, the region's best soils were farmed with increasing intensity, leaving few of the edge habitats upon which quail and other wildlife relied.[6]

New Deal activity did, however, provide an historical moment for a debate over agricultural practice, a debate that consumed the nation's community of natural resource professionals. The Dust Bowl in the southern plains states, rampant soil erosion in the South, and, ironically, overproduction of commodity staples like cotton, wheat, and rice, all led to a national reconsideration of agricultural practice. New Deal farm programs like the Agricultural Adjustment Administration and the Soil Erosion Service (later renamed the Soil Conservation Service) became involved—and in most cases were welcomed—in farmers' lives in ways previously unknown. The AAA paid farmers to participate in plow-ups and instituted allotment programs to reduce the supply of critical commodities, while the experts in the Soil Conservation Service scattered across the countryside to demonstrate soil-enhancing plowing and terracing techniques, cover cropping, woodland management, and more. At the same time, the federal government's increasingly complex administrative structure, especially as realized through its Land Utilization Program, expanded its control of land to include large swaths of former agricultural landscapes, and provided a critical organizational framework through which to practice wildlife management.[7]

Wildlife management, however, was a science very much in its youth. By the early 1930s it was still only practiced locally, with few concrete plans for regional or national application. Stoddard's work in the South, Leopold's in the Midwest, and the fledgling work of the SAAMI fellows were only just beginning to inform one another. Administratively, the Biological Survey remained the national clearinghouse for information on wildlife, but recommendations on how to manage land with a view for wildlife were local and spotty. As the recognized authority on southern wildlife and its habitat, Stoddard found himself in a unique position to influence policy. During the Quail Investigation he traveled widely in the South, advising landowners from North Carolina to Mississippi, and most states in between. His knowledge of coastal plain landscapes and their ecological processes was unrivaled at the time, but his frustration with what he considered the stifling labyrinth of government bureaucracy was equally incomparable. He continued to influence the implementation of wildlife management on both federal and state projects, but he was more interested in cultivating a private network of landowners and experts he felt was even more effective in restoring and maintaining wildlife habitats.

After returning to the Red Hills in early 1931, Stoddard helped to organize the Cooperative Quail Study Association (CQSA), a group of dues-paying private landowners spread throughout the Southeast. As the CQSA's only employee early on, Stoddard continued research on fire, volunteer and planted food sources, refuge cover, and fire ants; he consulted on membership landholdings; and he published annual and special reports based on research and field observation. The group had three classes of membership: 1) resident members owned land within a hundred miles of Sherwood and could call on Stoddard for on-site help at any time; 2) nonresident members, spread throughout the Southeast, received all publications and were entitled to a visit from Stoddard at least twice per year; and 3) corresponding members received the association's publications. The CQSA began with twenty-one resident members in 1931, and by 1935 had added thirty-four nonresident and corresponding members.[8] These are relatively small numbers, but the spatial breadth and temporal depth of the work made it one of the most substantial and influential wildlife organizations of its time. Stoddard estimated in 1935 that "now some sort of management is practiced on nearly a half million acres belonging to members of our Cooperative Quail Study Association. While admittedly much of it is crude, we have been working hard at it for over eleven years. As far as my observations go, this is the only portion of the country where quail management has been practiced *long enough*, through good seasons and bad, to get much of a line on results."[9]

It should be self-evident at this point that a concern for wildlife habitat meant dabbling in many areas of expertise. A quail covey's range, for instance, was not cordoned off at the forest's edge. They went where their habitat took them, which included the many fields and edge environments that covered the South. With that in mind, Stoddard had to look beyond the questions of forest, fire, predator, and prey to examine the history and management of the South's peculiar system of agriculture. *The Bobwhite Quail*, along with Stoddard's subsequent work, not only helps us understand the national reconsideration of fire and predators, then, but it also provides indispensable documentation of the South's early twentieth-century ecocultural landscapes.[10] Some of his most important findings were based on the assumption that the environment could not be treated outside of its social, cultural, and economic context. Of quail and their living conditions, Stoddard wrote, "It is becoming a difficult matter in the Eastern United States to find areas where quail are living under natural conditions, unaffected by man and his works."[11] Quail and their habitat were contingent on the qualitative actions of humans on the land, and Stoddard's work sought to direct those actions as much as possible.

In fact, conservation on the Red Hills preserves was as much about managing people as it was the environment. Tenants and their families not only remained on the quail preserves, they continued working the land in much the same fashion as they had since the end of Reconstruction. Stoddard's quail study altered some of those older activities in favor of wildlife, but tenantry formed the basis for many of his management techniques. Stoddard, in other words, was not simply advocating the conservation of a natural environment; he was among the first to call for the preservation and maintenance of the biological diversity found in this particular cultural landscape.

Stoddard adopted many of the land management techniques of tenant agriculture but remained aloof to the system's social dimensions. He never spoke out on the South's racial or economic inequalities, and he only expressed written concern about social issues when demagogues such as Georgia's Eugene Talmadge resorted to blatant attempts at race baiting.[12] Indeed, just as the preserve owners did in operating their preserves, Stoddard structured many of his conservation measures to depend on the organization of land and labor in the South. This dependence was partly ecologically based. The landscape mosaic he sought to mimic—one of small fields, woodlands, and edge effects—was a product of small-scale farming, and in the South that meant tenant agriculture. Though uninterested in questioning the human aspects of tenantry, his adoption of its landscape for conservation purposes mounted a vigorous challenge to the national move toward mechanized, clean agriculture. He reached out to preserve owners, farmers, and government resource managers, hoping to reverse what many observers and participants considered as progressive steps toward modern agricultural efficiency. Stoddard was mainly wary of their ecological effects:

> As the "red hills" are mostly good agricultural ground, cultivation in some sections here, as elsewhere, has become too intensive for quail. If more cover were left between the fields, even the most intensively farmed sections could continue to produce surplus quail. The tendency, however, to clean up all sheltering thicket cover with a view to destroying possible hibernating places of the cotton boll weevil and creating an appearance of neatness is proving disastrous to quail. Thousands of acres are classed as "shot out" by the misinformed, where bobwhites could not exist under any

> system of protection or restocking, simply because the environment no longer suits their requirements.[13]

For Stoddard, agriculture did not necessarily mean ecological barrenness; in fact, if farms were managed in certain ways, he believed they could actually increase plant and animal diversity.

Stoddard was not alone in his critique of the changing farmscape. Aldo Leopold was coming to similar conclusions, and Stoddard found it thrilling to find such a like, and receptive, mind when they met in 1928. Though much is written of Leopold's views on wilderness, and his role as an intellectual founder of modern environmentalism, much of his thinking was more contextually grounded in agricultural landscapes.[14] His appointment to the SAAMI survey forced him to look beyond the condition of wilderness lands and think seriously about the practice of agriculture on settled land. Stoddard's experience in the South, then, was a crucial source of information for Leopold about game on agricultural landscapes. Indeed, Stoddard and Leopold reinforced each other's views on agriculture and its place in the natural world. They joined together to point out that some of the gravest threats to wildlife and its habitat during the 1930s were occurring on the nation's agricultural lands, this at a moment when most conservation attention was either focused on the nation's public domain and its various management regimes, or on how to generate greater and more efficient production from agriculture.

Both Leopold and Stoddard thought about agricultural land as the key to maintaining the nation's natural resources, and farmers as its crucial caretakers. Their correspondence reveals much of their thought on the subject, as well as their frustration with farmers and experts alike who did not understand the urgency—and simplicity—of what they proposed. Their concerns, though rooted in ideas about the inherent value of nature, were also premised on several political, economic, and environmental realities that threatened the public's access to game animals. The basis of one problem was the apparent legal contradiction of the American "open" system of shooting: wildlife belonged to the state, but land—that is, wildlife habitat—was in possession of private individuals. The state could regulate hunting seasons and bag limits, but it had little authority, or political will, to address the core problem of diminishing habitat. The solution, if there was one, was to convince state game commissions and private landowners that disappearing game was directly related to disappearing habitat, not overshooting or predator activity.

Both Stoddard and Leopold thought a general dissemination of simple habitat restoration techniques would help. In discussing Leopold's Midwest, Stoddard related his experience in the agricultural South:

> In fine quail country everywhere we see many farms that are entirely unproductive of game due to agricultural practices that are adverse. In such cases they would, without doubt come up to the surrounding high level if food and cover were restored. . . . It should cost very little to let cover restore itself over a period of years on the average northern farm, and any quail increase secured by such means is surely worthwhile and economical. If only the grazing of woodlots, grubbing out of hedges and cleaning up of roadsides could be stopped and cover conditions allowed to correct themselves the increase in farm game would surely be striking.[15]

He was not asking small farmers to go to elaborate lengths to increase the game supply—he fully recognized their need to make a living off the land. Like Leopold, though, he thought the increase of game on private land to be in the public interest, and there were very simple, cheap measures landowners could take toward that end.

Nonetheless, the production of game did not operate within the normal parameters of supply and demand economics. As Leopold wrote in his landmark "Report to the American Game Conference on an American Game Policy" in 1930, "game is not a primary crop, but a secondary by-product of farm and forest lands, obtainable only when farming and forestry cropping methods are suitably modified in favor of the game. Economic forces must act through these primary land uses, rather than directly."[16] The other applied sciences like agriculture and forestry had a clear economic motive, and thus plenty of institutional support. The economic motive for game management was more nebulous, and the trick was to tweak the methods of resource production in favor of game, while also creating an economic justification for the tweaking. The task fell to a diverse coalition of interested parties, including biologists, nature lovers, sportsmen, and arms and ammunition manufacturers. By the late 1920s most of these interests converged at the annual American Game Conference, a meeting begun in 1911 by a consortium of gun companies, which quickly became an assemblage of the leading voices in American conservation. By 1928 the attendees recognized the work of Stoddard and Leopold, among others, as heralding a new approach to wildlife conservation, and they chose Leopold to lead a committee to outline a national game policy. Two years later, Leopold delivered a report his biographer calls "the most far-reaching document yet put forth by conservationists concerned with the fate of American wildlife."[17] Leopold addressed several facets of wildlife conservation, including the need to rely on sound research and experimentation rather than entrenched assumptions. Perhaps one of his most important points was that it mattered little what conference participants said about conservation until they included farmers and other landowners in the conversation. Without private landowners on board, everyone would lose out—the public; those invested professionally, economically, and recreationally in wildlife; and the wildlife itself.

In drafting the report, Leopold beat back and co-opted several potential policy approaches. The most threatening, in his mind, was based on a European system of game ownership and artificial propagation, which was backed most vocally by the sportsmen's group More Game Birds in America. Such a system, wherein the landowner also owned the game animals found within property borders, might increase the market value of game, but it would also make the regulation of game-related activities difficult, and the state protection of the game itself near impossible. A further complication was animal behavior—animals simply did not respect property boundaries. Though he dismissed the European system, Leopold did, along with several others, recognize the need to alter the American free system of shooting. As he explained in the report, "even if the system still prevalent in most states were effective in producing a game crop, it is increasingly ineffective in maintaining free public hunting on farms, because as hunters increase, trespass becomes a nuisance, and posting follows. Closed seasons, posting, or both, are the inevitable result on farm lands."[18] Why would a landowner want to increase game populations on his land, Leopold seemed to ask, only to encourage trespassing and necessitate posting? Stoddard asked similar questions, and thought that "in some quarters an abundance of farm game has come to be regarded as a liability rather than an asset, an unhealthy state of affairs from all standpoints."[19] W. L. McAtee was even more blunt in his assessment: "I think public shooting even now is dead in this country and the greatest trouble is that the sporting fraternity doesn't know enough to decently bury the corpse."[20] The landowner's right to close off game habitat already superseded the public's right to access game, and as sparsely settled territory diminished and trespassing became a growing problem, both wildlife and its public felt the squeeze. Leopold, Stoddard, and many other observers favored, as Leopold put it, "commercializing the shooting privilege *but not the game*, thus getting the advantage of private production incentive, without losing the advantage of state ownership and supervision."[21] Farmers could lease their land to hunters for a set fee, thus providing the motivation to diversify farm environments.

If Stoddard and Leopold did not fully approve of European laws regarding game, they did look across the pond for intellectual kinship. The English ecologist Charles Elton was an especially important contact. Leopold's relationship with Elton is well known—they met in July 1931 at the Matamek Conference in Labrador, Canada, a landmark conference on biological cycles, and became fast friends. Elton had published his book *Animal Ecology* in 1927, which would become a classic of modern ecology and a tremendous influence on Leopold.[22] At Leopold's urging, Stoddard and Elton became familiar with each other's work as well. Leopold suggested to Elton in 1931 that he have the "intellectual luxury of reading Stoddard," and also sent a copy of *Animal Ecology* to Stoddard around the same time.[23] Stoddard thought that Elton's work came "nearest of being the sort of ecology I can appreciate," a characteristic dig at theoretical work but a clear appreciation of its author.[24] Stoddard even had a chance to seek Elton's company during a 1935 tour of English and French game preserves. The two spent a day together at Elton's Bureau of Animal Population at Oxford University, and Stoddard enjoyed a week afield with Elton's assistant, A. D. Middleton.[25]

Much about the trip reinforced Stoddard's hope for agricultural landscapes. The English "hedge banks"—elevated hedgerows that divided fields—were particularly notable edge environments that Stoddard appreciated; indeed, Elton often argued that the English hedgerow was an ideal site for conservation.[26] Elton and Middleton were not land managers, however, and Stoddard saw very little of the type of management that was developing in the United States. In fact, as impressed as Stoddard was with Elton and Middleton's research, he was less enamored of the management methods he witnessed on European game preserves. He wrote to Leopold from France, saying that the hunting lands were "really most unusual. . . . I have seen almost nothing of land manipulation for game. It's all 'vermin' control, artificial feeding, and artificial rearing. . . . They know the partridges by their first names, and the best keepers can about lift them off their eggs."[27] Stoddard, Leopold, and other American wildlife managers had spent years by this time attempting to move beyond such approaches. Like Elton, they were making connections between habitat, food supply, and population fluctuations, and they were also learning how to shape environments to meet their management goals. And raising and releasing game animals was of diminishing value to the American wildlife agenda. When Stoddard arrived home after his six-week tour of the European countryside, he was never more sure of the American approach to wildlife management.

Despite the interest among U.S. wildlife managers in shaping agricultural landscapes, farmers themselves had little information about building and sustaining wildlife populations. Stoddard's work was known in relatively circumscribed circles of wildlife experts and professional land managers like those in the Red Hills, as were the SAAMI fellowship projects. But ordinary farmers were far out of the loop. Coming on the heels of Leopold's charge to conduct research and disseminate results throughout the country, the Biological Survey actively sought a remedy through the familiar public-private administrative model established by the Cooperative Quail Investigation and SAAMI projects. The most ambitious venture, spearheaded by the DuPont De Nemours & Company, was to establish a series of game management demonstration projects throughout the Southeast. The original plan, outlined by DuPont researchers A. C. Heyward and Henry Davis in the fall of 1931, designated twelve research and demonstration areas in eight southeastern states to be administered cooperatively by the Biological Survey, the American Game Association, and local or state sponsors. They were to last three years. The research areas, known as "type I" projects, would duplicate Stoddard's Red Hills work, and investigate any modifications "made necessary by changing terrain and climatic conditions," according to the original proposal. Demonstration areas, or "type II" projects, were to be applied manifestations of type I projects, "so that these findings may prove of the greatest practical value to the general public."[28] In his capacity as independent consultant to the survey, Stoddard gave instruction to researchers and land managers, and he visited each site at least twice per year.

The program was beset with problems from the beginning. Funding relied on local sponsors, often a mix of politically connected sportsmen and state game commissions; and each project's fate depended largely on their interests and goals, which often did not correspond with the program's original purpose. The disintegration of the Georgia project in LaGrange was fairly typical. When the American Game Association alerted state contacts of the budding program in late 1931, Georgia's commissioner of Game and Fish, Peter Twitty, expressed immediate interest. Twitty already knew the local American Legion in LaGrange had an interest in increasing west Georgia quail populations, and they had access to about a thousand acres Twitty thought ideal for either type of project. The Biological Survey sent Stoddard to scout the land, meet with Twitty and the local sponsors, and generally size up what looked to be a solid prospect in January 1932. The land itself looked like a good representative of the area's environmental conditions—part operational farm, part eroded old fields, and part cutover timber.

Stoddard considered it a good spot to demonstrate wildlife restoration techniques in a heavily agricultural region, but he did not get a good vibe from the initial meeting with local sponsors. Locals seemed to think the demonstration area would act as nothing more than a quail propagating plant. As he wrote in his report to the survey, which was shared with all interested parties, "I understand from remarks passed during our meetings that local financial aid was being secured because of the likelihood that a sufficient number of birds could be produced for restocking other areas to justify it. This is not my understanding of the purpose of these Demonstrations." He reminded Twitty that "the actual benefit to the sponsors should consist of the information secured as to how to get birds economically over the vast areas of similar type in their region; not from the birds actually produced from the test areas."[29] Twitty soon pulled his support, and the project fizzled.

Similar sponsorship problems plagued the other sites. As Biological Survey agents Wallace Grange and Ross Stevens reported later that year on the program in general, "the difficulties arising are with the human rather than the environmental elements."[30] Some sponsors could not abide letting hawks fly overhead without a shot; others would not allow the experimental use of fire; and still others simply wanted to propagate birds and restock depleted lands. Even the sponsors of the Oklahoma project, one of the few to complete the full three-year contract, considered "the project only as a production farm" to supply a surrounding hunting club, according to lead technician Verne Davison.[31] The new approach to wildlife management—restoring habitat rather than restocking populations and exterminating predators—was clearly not gaining much ground in popular circles, even if they were interested in the subject.

There was much work ahead to ensure that public policy and local wildlife projects reflected current research. Stoddard, for his part, continued important research and experimented endlessly with management techniques on land over which he had some control, but he gradually decided to withdraw from the public fold. Of the early wildlife professionals, Leopold became the point man on pushing the public toward a more biocentric view of nature. Throughout the 1930s he worked hard to build his now famous ethical argument for land conservation, one that extended ethics beyond human relationships to include their relationship with the environment. In two essays in particular, "The Conservation Ethic" (1933) and "Conservation Economics" (1934) Leopold argued that both the government and individual landowners alike had an ethical duty to conserve land using all the scientific means at their disposal, whether in the soil, forest, ecological, or wildlife sciences.[32] Stoddard read both manuscripts, and of the latter wrote that it "strikes me as a very keen paper and very much the meat of the cocoanut. I only wish that a copy could be sent to every conservationist, farmer, and citizen of the country; it should help to raise the citizenship standard."[33] Stoddard recognized Leopold's skills with the pen, as well as his gift for synthesizing complex social and environmental problems.

Stoddard also accepted his own limitations as a critic. When Leopold encouraged him to join the policy debate more vigorously, noting that "you are better entitled to criticize the [wildlife conservation] field than anybody else in the country,"[34] Stoddard demurred: "I am so constituted that I am apt to assume that the other fellow is doing the best he can under the circumstances, and could not write a critical review of a section of the conservation field to save my life."[35] Considering his many criticisms of contemporary land management regimes, one might suspect this a tongue-in-cheek response, but in many ways Stoddard was content to let his work speak for itself. His strength was in employing his management techniques on lands with receptive guardians, not fighting policy battles. He was more than willing to let Leopold and others assume the public face of wildlife management. Stoddard shared with Leopold the hope that, with a bit more attention to the details of wildlife habitat and a bit less enthusiasm for modernization at all costs, the nation's farmers might continue to protect vital biotic reservoirs from becoming monocultural barrens. But he played his part by turning inward, to his quail preserves and their region, the American South.[36]

In Leopold's midwestern context, it was relatively uncomplicated to celebrate traditional nonintensive agriculture as protective of biological diversity. In Herbert Stoddard's region of interest, however, the farm and forest habitats of the coastal plain came packaged within a socioeconomic system that made such claims of ecological beneficence far more problematic. Not unlike the birth of many other conservation regimes, the history of southern quail management was rife with social inequality.[37] An examination of Thomas County, where Stoddard did the majority of his consulting work, well represents the prevailing agricultural and social conditions found in other coastal plain plantation districts. In 1930 Thomas County was very much rooted in the rural institutions of tenant agriculture. Its total population stood at 32,612, about 49 percent of whom were black. The black population previously held a majority, but their number fell by 1,407 in the 1920s due to migration north. With an urban population of 11,733, town life in Thomasville, Metcalf, and Boston, among other smaller crossroad communities, was bustling, but a wide majority of 20,879 still worked and resided in the countryside.[38]

Land-use trends in Thomas County also mirrored those in other plantation districts, though it was more heavily forested than most. Of the county's total 339,000 acres in 1930, 112,142 were in cropland and 13,142 acres lay idle or fallow. The average sized farm, including those of nonlandowning tenants, grew in the previous decade from 89 to 112 acres, reflecting the slow march toward consolidation. Pastured land was common, though it was not in the improved variety of grasses that spread across the southern landscape in later years. Of the 74,074 acres that were in pasture, over two-thirds was woodland pasture, which simply meant open range pinelands. In addition, another 73,863 acres were nonpastured woodland, most of which was found within quail preserve boundaries.[39]

Despite so much land being locked up by northern-based preserve owners, Thomas County ranked across the board as one Georgia's top producers of agricultural goods in 1930. It had the largest population of pigs—36,400—of any county in Georgia, and only five counties had more head of cattle. It ranked eighth in corn acreage with 48,935, and second in sweet potatoes with 2,169. Its cotton acreage was unremarkable in a cotton-dominated state, but 19,079 acres was substantial nonetheless. Thomasville's location on several active rail lines made it a regional market and transportation hub, leading to a substantial increase in truck crops throughout the county. Between 1920 and 1930 the total acreage for commercial vegetables increased from 1,936 to 7,995, and it became the second ranked county in both vegetable acreage and dollar value.[40]

Thomas County had a relatively diversified farmscape, but its tenure arrangements closely mirrored those of other southern regions. Tenantry and sharecropping pervaded Red Hills farming during Stoddard's initial quail investigation. The 1930 census counted well over half of total farms in Thomas County as tenant operated, with the twenty- to forty-acre lease being the most common arrangement by far.[41] Even when part of a large unit of ownership, these farms were small and scattered. Credit arrangements on the quail preserves continued to emulate those on locally owned lands throughout the

1930s, as did tenant farmers' reliance on cotton for cash. Several standard contracts prevailed. One Leon County extension agent reported of the quail preserves in 1937 that "these plantations operate on a fixed commodity tenant basis," likely a standard number of bales per forty acres. "Incidentally," he continued, "cotton is about the only cash crop grown by tenants and often does not produce sufficiently to give the tenant any profit above his rent."[42] Ichauway Plantation in the Albany region, on the other hand, used a typical sharecropping arrangement. Its 1939 "Farm Program" rules stated, "All produce [meaning crops], other than vegetables and meat raised for home consumption, will be divided equally between the tenant and ourselves. . . . All farm produce must be turned over to the Superintendent to be sold by him." Tenants supplied their own stock, implements, and seed, while Ichauway provided fertilizer. To keep tenant families afloat through the year, "advances to tenants will be made for seven months of the year by the Superintendent on the first Saturday after the first of each month."[43] As was typical in southern sharecropping and tenantry, to make it through the next year a tenant had to borrow on the next year's crop from the landowner or a merchant, thus making it near impossible to escape the crop-lien cycle.

Though Stoddard was not interested in critically assessing the region's attachment to tenantry, he was interested in directing the actions of those tenants who remained on quail lands. There was no shortage of potentially good quail land in the Depression-era South, and virtually all of it purchased for quail management had tenants spread throughout. On land with tenants, Stoddard wrote, "it is undoubtedly best to keep all who respond to fair treatment and cooperate with the owner in special matters."[44] Tenants continued most of their traditional patterns of land use, like cultivating small fields and gardens, and hunting small mammals, but the "special matters" to which Stoddard referred were not inconsequential. He made it clear that managing this environment would also involve managing the tenants remaining on its lands.

The most significant management changes recommended by the quail report were to eliminate free-ranging cattle, poultry, cats, and dogs. Cattle competed for quail food plants and trampled much of the nesting range, while free-ranging chickens may have transferred diseases to quail. Stoddard recognized that controlling tenant activity was a ticklish matter, and he provided special counsel on dealing with roaming cats and dogs: "As the greater portion of these animals belong to the tenants living on the land, tact and diplomacy rather than force have to be relied upon in handling the delicate problem of the restriction in number, control or disposal of these pests."[45] It would be many years before preserve owners rid the range of tenant-owned cattle, but four years into the study, preserve owners developed a standard agreement with most tenants concerning domestic pets and small mammal hunting. Stoddard held a series of meetings with tenants in 1928 to outline the preliminary results of the study, and "to get maximum cooperation from the tenants in efforts to build up the quail supply." He encouraged them to continue hunting possums, raccoons, and skunks as they had for generations, and "the tenants all agreed to give up all of their cats and all but one dog, and that to be tied up except during the time it was being used in handling stock, etc."[46] Clearly, not all tenant practice was beneficial from an ecological standpoint. Stoddard celebrated the unruliness of the tenant landscape, but he also sought to manage many of the actions that had helped produce it.

The landscape of tenantry was fraught with environmental paradoxes. It contained considerable ecological diversity, but it also created a very real potential for ecological destructiveness. The agricultural practices of southern tenantry, especially on hilly land, have long been associated with soil erosion and infertility.[47] Much like the piedmont sections to the north, the soil of the Red Hills gave way on countless hillsides, and was leached out on others. One of many consequences was a harmful effect on wildlife habitat. Indeed, Stoddard used part of the quail study to rail against cotton monoculture and its effects on the soil. "The methods used in cotton raising are highly detrimental to quail," he argued. "Not only are cotton fields an unfavorable quail environment, but the planting of the crop year after year in the same fields, without rotation, has put hundreds of thousands of acres into an unproductive condition."[48] He cautioned those interested in developing lands for quail to closely consider past land use, for worn out lands would take time to replenish. Land where "the fertility of the soil has been exhausted to a point where it can not produce a vigorous growth of weeds and leguminous plants will not support quail in abundance."[49] On land prone to erosion problems, Stoddard enthusiastically encouraged terracing. Not only could terraces "preserve hillsides from destructive erosion," they could also "be made to furnish ideal areas for quail cover. Their importance is so great both from the viewpoints of soil retention and as havens for quail and other bird life, that it is urged that they be put in all agricultural quail preserves in rolling country."[50] Such advice was not unique; agricultural specialists had advanced these cropping techniques in the South for several decades, and would continue to do so through the New Deal. Stoddard, though, was less interested in agricultural productivity than he was in biological productivity. He did recognize the presence of human communities on the land, however, and considered small-scale agriculture perfectly compatible with the region's natural communities—if it did not destroy the soil.

Terracing made headway on preserve land in the early 1930s, and many landowners and tenants even began to curtail their reliance on cotton. It remained the most important cash crop on most quail preserves until World War II, but some owners were on the lookout for alternatives, an effort as rooted in a concern for quail as it was in a concern for tenant well-being. As Stoddard worked out the administrative details of the CQSA in 1931, preserve owner Arthur Lapsley proposed the organization take on a dual role as quail research clearinghouse and farm cooperative. Reflecting on quail and tenant farming, he reminded Stoddard, "You know our 'renters' depend on cotton for their cash crop. Cotton is no good for quail, and worse for the land. If something could be found that was both good for quail and land—and could be *sold* for *cash*, it would be a very great help." Moving away from cotton would help all interested parties, and Lapsley considered a recent fallout in the cotton market an opportunity to diversify: "The low price of cotton, and the possibility that thru our 'Quail Club,' small quantities of a crop could be pooled, and thus sold to advantage encourages me a little. Let's try our best."[51] Stoddard suggested intercropping velvet beans with corn as a possible alternative. Velvet beans were good nitrogen fixers and used widely in the South for erosion control, forage, and household use; and corn, of course, was already a staple of both the home and market. Stoddard was unsure, however, "whether a good price can be secured on the market year after year . . . to keep the renters going."[52] Indeed, corn prices fluctuated like most staple commodities, and it was not much of a cash crop in the South like it was in other regions.[53] Nevertheless, it was becoming the foremost field crop on many preserves. By 1934 Ichauway mandated that each tenant plant a total of thirty-eight acres in field crops, with eight acres in cotton, ten in peanuts, and twenty in corn intermixed with velvet beans.[54] Dan Lilly, a tenant on Ichauway, remembered working "about more corn as anybody around. . . . So much corn everywhere. These little bird patches in the woods. . . . We planted them in corn to take care of the birds."[55] For Lilly and many other tenants, corn would increasingly need to take care of their families as well.

Beyond his hope for tact and diplomacy, Stoddard left little record of what he personally thought of the region's economic and racial disparities, but it is apparent that neither he nor the preserve owners sought to overturn deeply entrenched patterns of power in the South. Racial patronage and concerns over a dwindling labor supply were as common for the preserve-owning industrialists as they were for the southern planter classes. One preserve owner, for example, thought one of her servants "was so at one with nature that we could easily imagine him speaking the language of the little woods-creatures, and sharing their secrets."[56] This language of naturalization pervades much documentation when the conversation shifted to black labor. It is significant, though, that conservation on the quail preserves did not mean general expulsion from the land like it did in so many other contexts. Indeed, the partition and control of the quail preserves actually created a local, or more precisely a private, commons for the residents of the properties. The lands were posted to outsiders, to be sure; but, outside of the "special matters," those who lived and worked on the preserves had free range over the preserve environment.[57] Nevertheless, the revolutionary ecological and wildlife management developments on these quail lands were not accompanied by a revolution in social or racial values.

The tenant experience on a quail preserve varied from property to property, and it is difficult to gauge thoroughly with extant evidence. Most available oral histories of tenant life reveal a comfortable existence. Lucille Glenn Morris, of Susina Plantation, recalled life as "enriching, nurturing, and enjoyable. Love, brotherhood, industry, responsibility, and goal-setting were some of the values instilled in a lovable, family-like setting."[58] In a time of scarce farm income, and even scarcer employment, many African Americans in the Red Hills coveted the security the preserves provided. Morris thought that "many of the surrounding plantations were blessings for rural blacks that had limited outlets for gainful employment in South Georgia during the early 20s."[59] Frank Delaney of Pebble Hill plantation thought that the poor living and working conditions of nonpreserve land created a great demand for access to farm land on the quail preserves. Because "you had people living on the plantation that was having it a whole lot better . . . there were people trying to move on the plantation. We had to have a system to screen people on the plantation."[60] Such demand likely had as much to do with land shortages and poor conditions elsewhere as benevolence on the part of preserve owners, but there did develop a sense of pride in being affiliated with such a prestigious landscape.

Even still, tenants worked and lived with strict regulations, a private governance of privileges, not rights. Job responsibilities on the preserves varied widely. Some families simply rented land on shares and farmed according to their own inclinations. Others farmed during the growing season and worked wage jobs for the landowners during the winter, ranging from cutting lumber to working dogs and driving wagons on quail hunts. Another group of residents worked full time for landowners. Full-time women laborers typically cooked and cleaned, and men worked the grounds, the dog kennels, or as chauffeurs. On Pebble Hill, perhaps the most progressive preserve in terms of tenant treatment by the 1930s, tenant families paid no board, had free access to a full-time nurse, and could even tap into a college scholarship fund established by longtime owner Kate Hanna Ireland and her daughter, Elizabeth "Pansy" Poe.[61] Frank Delaney recalled ample resources to draw on for all Pebble Hill residents: "There was enough land allotted to each family, each household. If you wanted to plant a garden that was an acre, no problem . . . from corn to carrots, beets, strawberries, anything you wanted to plant. . . . Every household had a garden, a chicken house, a smokehouse, and a garage. My dad used to plant a few acres of corn; we kept a big garden. We . . . had hogs, and the neighbors had hogs. . . . Whatever you had, you just shared it, you know."[62] Memories of life on other preserves echo those of Delaney. Irene Hudson, a lifelong resident of Ichauway Plantation south of Albany, Georgia, called up the 1930s as a time of subsistence living on little cash, though with more nostalgia than bitterness: "People then didn't hardly buy, they didn't have to buy that much stuff. See, they were farming, and they raised the cows, they raised their own hogs, so they had to buy tobacco, sugar, flour. They raised corn, they take their corn and shell it, put it in a sack, carry it to the grist mill, have it ground, you didn't have to buy no meal, you didn't have to buy nothing but flour and sugar . . . we didn't eat nothing but cornbread, peas, and all different kinds of vegetables, sweet potatoes."[63] Indeed, Ichauway required that "every tenant, without any exception, must have a good garden and grow enough vegetables for home consumption."[64]

Such recollections put one in an idyllic frame of mind, and those from Pebble Hill even allude to a burgeoning black agrarian middle class. But tenant conditions were less than consistent regionwide. Leon County's extension agent expressed repeated frustration throughout the 1930s that the quail preserves did not allow tenants to modernize their farming techniques. His 1938 report noted, "Most of the 914 Negro tenants, accounting for fifty-five per cent of the farms of the county, are tenants on plantations. None of the plantations allow fences, so livestock raising is impossible. Farming equipment is limited and inadequate. Workstock is usually one mule or horse, or two oxen. The average value of equipment per farm in the county is only $125, and the Negro tenants will not average more than $50."[65] Similar conditions prevailed elsewhere. West of the Red Hills, where newer quail preserves began to flourish on the Dougherty Plain region near Albany, Georgia, tenants remembered a life of hard work and little economic mobility. Dan Lilly farmed on shares before World War II and remembered many tenants harvesting corn, peanuts, or cotton only to end the year indebted to the plantation store. After reconciling the books for seed, fertilizer, and household goods, "you'd have to start next year with that debt on you and another one coming. That's why you wrapped up in debt . . . they'd never get out."[66] Some amenities of the quail preserves offered a slight buffer against market realities, but even so, debt was debt.

Tenants were clearly caught in a debilitating system, but they did possess a great deal of knowledge about the environments in which they worked and lived. And they used it to their advantage time and again. Resident quail preserve tenants had a particularly important, though largely anonymous, role in Stoddard's wildlife work. In the spring and summer they located and helped to monitor nesting sites, and they often acted as guides in unfamiliar territory throughout the year. When preserve owners agreed to pay tenants for locating quail nests and handing over targeted predators, they quickly seized the opportunity to add to their meager cash flow. During the summer months Stoddard spent day after day "visiting quail nests previously located and going here and there with negroes to see new ones."[67] Though he noted that "many tenants and field hands normally find many quail nests in the course of a season," Stoddard also assured preserve owners that a fee of fifty cents per nest "stimulated a lot of special search."[68] In addition to locating nests, tenants also played a crucial role in Stoddard's study of quail predators. As discussed in chapter 4, they turned in small mammals and reptiles for stomach analysis, and they reported any predation activity on quail nests. Tenants, in effect, became ad hoc scientific workers while carrying out their many other tasks.

Despite the many social and economic inequities, Stoddard advocated strongly for the landscape of tenantry as it developed on the quail preserves, and thought it offered an ideal biotic refuge. That landscape was quickly changing regionwide, though, and not in favor of wildlife resources. When Georgia's state game commissioner, Peter Twitty, asked Stoddard in 1932 to distill *The Bobwhite Quail* into a pamphlet more readily available to farmers, he outlined a landscape already familiar to any southern farmer of long experience. He wrote of successional seed plants—more likely known as weeds to progressive farmers—that grew vigorously in the highly disturbed agricultural landscape. Ragweed, beggarweed, pigweed, rough button-weed, Mexican clover, and bull grass among many others sprouted after a field was "laid by," and occupied field edges and open woodlands throughout the growing season. A variety of tree and shrub masts also contributed to make the landscape a quail haven. Stoddard advised farmers that "quail are fond of, and more or less dependent upon a wide variety of small wild fruits, and the 'mast' from trees for their living," especially wild black cherry, dewberries, sassafras, blackberries, wild plum, huckleberries, and mulberries, "which are often abundant on not too intensively cultivated farm lands." It was also "well to remember their value to quail, wild turkeys and other birds when considering the cutting of wild cherry for fence posts, and brushing out around fields and along fence lines, roadsides and so forth, for the destruction of such food and shelter producing vegetation may be the means of reducing the number of quail on the farm."[69]

The other key to the functional quail landscape was cover. Many of the same plants that provided food also gave quick cover from predators; "thickets and vine tangles around field borders, on fence lines and roadsides, and here and there in open woodlands" were essential requirements for quail. Coverts not only aided quail, but as discussed previously, they substituted for predator eradication. Again, the landscape of tenantry had most of these measures built in, but "where farmers cut out such refuge cover to give their farms an air of 'neatness,' . . . a decline in the numbers of quail and other thicket-loving birds is inevitable."[70] Such neatness threatened the traditional wildness of southern farms, and on lands outside of his control, there was little Stoddard could do about the trend beyond recommending otherwise. On the quail preserves, however, his word carried a great deal of weight, and much of his advice was carried out to the letter.

In the nomenclature of the time, Stoddard called tenantry a "primitive" or "crude" system of agriculture, but he was unequivocal about its ecological superiority to modern systems. In his chapter in *The Bobwhite Quail* titled "Preserve Development and Management," he advised that the best "situations in which to establish new preserves, consist of ground of medium or low price, land that is, or has recently been, under a system of crude agriculture." "*Where fields are small and well distributed*," he continued, "with small open woodlands between, and thicket cover is plentiful, the all-important matter of environment is favorable to start with and the food supply in most cases can be built up quickly."[71] His preference for small-scale agriculture was practical; it had little to do with a romantic primitivism in the face of modernization. Nor did it have much to do with the preservation of untouched nature. Stoddard's advocacy was for the active management of cultural landscapes to produce abundant and diverse wildlife populations.

It was here, in the suggestion that species diversity in the longleaf environment was contingent on landscape diversity, that the influence of tenant agriculture made its biggest mark on Stoddard's ideas about ecological land management. The greater the number of landscape types spread throughout the countryside, he proposed, the better chance there was of promoting a wide variety of plants and animals. In regard to quail, he explained, "the necessity for providing feeding grounds *in close proximity to thicket cover* explains why very diversified areas 'carry' more quail than do either great fields or large solid blocks of woodland."[72] In Stoddard's mind, there was a certain advantage on lands as heavily worked as the Red Hills—or indeed, the whole of the eastern United States. The key was to spread this patchwork landscape evenly across the countryside. Stoddard believed the goal of any landowner hoping to realize wildlife conservation "should be to create *maximum diversification*, and have small woodlands, small fields in crop and fallow, and roosting and nesting grounds evenly distributed over the terrain."[73] It was not that Stoddard did not appreciate vast stretches of undisturbed wilderness—he did. But in these agricultural landscapes of the longleaf pine woodlands, wilderness was not the issue. Indeed, if left to the untended devices of nature, without fire or other human regimes of disturbance, wildlife diversity would almost certainly decline.

The issue, for Stoddard, was to spare wildness and wildlife on a middle ground between the old unmanaged methods of wilderness preservation and an increasingly mechanistic approach to growing crops. The key to wildlife management in this region, he concluded, "consists in diversifying the vegetation as much as possible and providing a balance of open woodlands, weedy fields, cultivated and fallow ground, thickets, and scattered grass or broomsedge areas of proper density and small extent."[74] Instead of attempting to impose nonhuman nature on a very human landscape, Stoddard took this southern environment on its own terms—and its terms hinged squarely on ecological disturbance. It was a plowed, chopped, grazed, ditched, and burned environment of dynamic human and nonhuman activity, and Stoddard was looking for ways to best manage, or order, such disturbance. Wildlife abundance literally depended on it.

Figure 10. Aerial Photograph of Melrose Plantation, Thomas County, 1928. Courtesy of Tall Timbers Research Station.

The landscape effect of all this agricultural activity was a complex tapestry of variegated environmental space that is perhaps best explored in an image. Figure 10 is a striking 1928 aerial view of a portion of Melrose Plantation in Thomas County, one of several Hanna family properties in the Red Hills. It is perhaps the best visual example of Stoddard's ideal forested landscape. Melrose was approximately four thousand acres (a small slice of the Hannas' total Red Hills acreage), with "a large aggregate acreage in corn, often well-distributed in small fields," according to a 1933 report on the property.[75] The photo shows what today's conservation biologists might call landscape corridors on a small scale. Small fields break up large blocks of woodland, which narrow to form bottlenecks from one block to another. Woodland animals, then, had large areas in which to roam, as did species favoring edge habitats. Down on the surface, beneath the forested overstory, fire created a diverse vegetative understory with plenty of food for seed-eating animals. If we could continue to zoom out to cover the entire quail preserve region, we would see a similar, though certainly not identical, pattern of landscape diversity. The point is that the region was far from an ecological monolith; it was wholly covered neither by fields nor by forests. It encompassed a diverse patchwork of environments created from human land use. This, in other words, was a landscape of intense human activity, but it is clearly not a landscape of industrial activity.

Such landscapes as that of Melrose, however, were quickly being replaced by the homogeneity of industrial agriculture and forestry. Decreasing landscape diversity, in fact, was Stoddard's principal criticism of southern land-use trends regionwide. While the expansion and mechanization of agriculture was the culprit on land with good soils, a companion development, the total abandonment of agriculture and its management techniques, was equally alarming on poorer lands. This became particularly evident on newly purchased government lands, which provide an instructive contrast to management on the Red Hills preserves. Beginning in 1934, the New Deal's Land Utilization Program began purchasing marginal and submarginal land throughout the nation, and marked the cotton-growing and cutover lands of the South as one important problem spot. Administered at various times by the Federal Emergency Relief Administration, the Agricultural Adjustment Administration, the Resettlement Administration, the Bureau of Agricultural Economics, and the Soil Conservation Service, the Land Utilization Program had as its goal to take poor land out of production in an effort to correct the widespread reality of farm poverty. In a few short years the Land Utilization

Program purchased 11 million acres of mostly worn-out agricultural lands, relocated thousands of people, and transferred management of the land to various federal bureaus or to state governments.[76] In the Southeast, much of the land eventually became national forests, managed by the Forest Service, or wildlife refuges, managed by the Biological Survey and later the Fish and Wildlife Service. Most observers interested in wildlife protection considered this new public ownership of resources a boon to the movement, and it surely was in the long run. But there were also limitations to the effectiveness of such an approach. As previously noted, the government could take only so much land out of private ownership. In Stoddard's estimation, though, the most severe limitation rested in the ability to properly manage what they had.

In these newly created national forests and wildlife refuges, the vernacular landscapes of local agricultural practice gave way under the more centralized approach of the federal government. Individual decisions once based on living within a landscape now came by way of administrative decree. In a series of inspection reports he made during the mid-to-late 1930s for both the Forest Service and the Biological Survey, Stoddard cautioned those who thought good conservation on "sub-marginal" agricultural land would come about by simply dispensing with people and allowing a "reversal to nature."[77] His greatest concern was not the loss of agriculture, per se—much of this land was, indeed, worn out and could do without the plow for a time. Rather, the greatest absence on the new federal lands was the landscape effect that older agricultural practices created. Active wildlife management, he argued, could mimic these agricultural practices yet avoid their debilitating effects on the soil.

Stoddard's reports to the Forest Service and Biological Survey contained a consistent theme: do not expect to produce desirable wildlife habitat—and by extension, recreational and aesthetic landscapes—by simply planting trees. As discussed in chapter 3, the rhetoric of southern reforestation took off around the turn of the twentieth century, and by the 1930s had reached the level of dogma. Much of the southern coastal plain was in deplorable condition, so there were legitimate economic, as well as ecological, reasons for talk of reforestation. But, again, the desired environmental conditions of New Deal planners and foresters were not those of wildlife managers like Stoddard.

On what would become the Chickasawhay National Forest in southern Mississippi, for instance, Stoddard advised, "the provision of well distributed and adequate openings in otherwise solid forest is of the greatest importance to all the game species," and "open park like areas of pineland, kept so by the periodic use of controlled fire, are very desirable supplement to the open fields."[78] Small fields interspersed among open pine forests was the very core of the agricultural landscape of his home in the Red Hills, but Stoddard knew very well that his management goals came into conflict with those of foresters. In his report on the future Wambaw, Pisgah, Cherokee, Nantahala, and Chattahoochee National Forests, he warned Forest Service officials that "many of the questions discussed so frankly are highly controversial in nature. While I have considerable confidence that the opinions and suggestions advanced are sound from a game management standpoint, I realize fully that they may conflict with forestry or grazing interests."[79] The restoration ideal of government foresters was driven by market economics, which meant fast-growing, densely populated stands of merchantable trees.

What most rankled Stoddard about management on these government lands was the replacement of an ecological tapestry he had worked so hard to decipher with a previously unknown ecological community that had few ties to the historical environment. On the southern coastal plain lands, in particular, where longleaf pine woodlands were greatly denuded but resilient, and agriculture had been relatively low-level and scattered, Stoddard thought federal agencies were too hasty to act before they knew much of anything about coastal plain ecology. His report on the Chickasawhay continued rather strikingly:

> A policy decidedly open to criticism from a wildlife standpoint in my opinion, is the planting of large solid blocks of pine all over the uplands, without leaving from 10% to 15% of well dispersed openings so essential to the welfare of wildlife. The planting of Slash Pine particularly would seem to the writer questionable from more than the wildlife standpoint.
>
> It would seem that the pine plantings could be made to fit into the picture better from a wildlife and recreation standpoint were the slash planted adjoining and parallel to the creek valleys, draws, and lowland areas where the hardwoods hold forth. . . . This would leave the uplands for the long-leaf, where this tree naturally grows, and this type could be periodically control burned for both silvicultural and game management purposes. . . . Such a development would have a high scenic and wildlife value, and probably grow very nearly as much timber as any other."[80]

Stoddard concluded his summation on the Mississippi lands, writing, "Thousands of acres of solid slash pine, without a proper balance of open lands, will ultimately be barren of desirable wildlife."[81] Stoddard's idea of desirable wildlife included most everything—the birds, mammals, reptiles, amphibians, and insects that made the longleaf pine woodlands whole. Indeed, these recommendations look very similar to today's ecological restoration efforts on longleaf lands. Stoddard maintained that fire-intolerant pines had no business on the coastal plain uplands, where both natural and human-set fire had shaped the longleaf pine woodland habitat for millennia; only longleaf pines could begin to restore these lands to any semblance of ecological function. But Stoddard's recommendations for the southern national forests also derived from the agricultural lands he studied in the Red Hills. By removing people from the landscape, the federal government effectively removed a type of landscape diversity endemic to small-scale agriculture on the southern coastal plain. Most importantly, the common-sense use of fire disappeared, and along with it went the landscape tapestry of small fields and open forest. Without small-scale agriculture and its practitioners spread throughout these vast tracts of land, or at least a deliberate mimicry of it, dense forest cover replaced diversity.

And so Stoddard created a hybrid form of conservation that stood in opposition to the clean-slate mentality of modern agriculture and the wholesale agricultural abandonment of the federal government. What we see through Stoddard's applied science is a growing realization that some environments should be managed locally; part of the problem was the scale of management and the knowledge of the managers. In addition, Stoddard argued that some land management should look beyond the economic use of resources to address ecological abundance as well. Beyond the lands of the nation's most wealthy recreationalists, the best opportunity to do so was on government-owned lands; they lay outside the trappings of the market, and were taken out of production for the stated goal of rehabilitation and restoration. The ecological knowledge to encourage species diversity existed, and Stoddard considered the management steps he took in the Red Hills to be a perfect fit for government lands. But he and others also knew the Forest Service was not the best agency through which to angle for diversity; their purpose was reforestation and timber production, a plan that did not fully square with wildlife management.

The Biological Survey and their Wildlife Refuges, though, had few such hurdles. The Wildlife Refuges were just that, places devoted "to the use for which they were primarily intended by Nature."[82] Here was an opportunity, thought many in the wildlife community, to publicly showcase the recent strides in wildlife management. When Waldo McAtee informed Stoddard of a 1934 purchase to be managed by the Biological Survey, Stoddard encouraged him by relating his own experience managing large tracts of land:

> Am glad to hear that you have a tract of land to develop, and you will have lots of fun doing it. Now I take just as much pleasure developing lands primarily for quail and turkey, regardless of the shooting, and for this reason: Everything we do to improve conditions for game improves conditions for all wild life of similar requirements, as well as the game. The net result is that we have vastly more game than when we started, as well as dozens of other forms of wildlife. The millionaire game preserve owners are doing more to increase seed eating wild creatures than any other element, though few seem to realize the fact. . . . I would suggest that you try all of the stunts we are trying here for food and cover improvement, as well as what you can think up yourself.[83]

In developing land for game animals, then, Stoddard surreptitiously carried out a program for nongame species as well. His persistent problem in expanding that program to more land, however, was in the suspicion by some officials that the preserve owners had unlimited coffers from which to draw, thus making such management unfeasible on most landscapes.

Indeed, the preserve owners spent a great deal of money on their exclusive landscapes of retreat. But despite the extravagance, Stoddard felt the land management measures could be easily and cheaply replicated on government lands. McAtee, however, expressed skepticism in drafting an internal document outlining the Biological Survey's role in administering public land. After identifying the CQI as the foundational work in wildlife conservation, he touched a nerve in Stoddard by writing, "the private preserves of the wealthy are certainly not the place to look for guidance as to economical game management."[84] Though he was sometimes critical of the lifestyles of the preserve owners, Stoddard was forever a defender of their landscapes.[85] He quickly identified a contradiction in McAtee's essay, writing that the methods developed in the Red Hills were "being lifted bodily by all the upland game development projects in the country," so the preserves were, indeed, the place to look. He went further, though, making clear that extensive management did not necessarily equate with profligate spending:

> As to making management pay for itself, we have a constantly increasing number of members who are becoming interested in this. . . . Several

> are trying with best available advice to make their woodlands productive, regulate their agriculture so as to be as profitable as possible as well as ideal for quail, doves and turkeys, and otherwise make their places as nearly self-sustaining as possible. The high costs you mention per quail, duck, or other game birds does prevail on some places, but seldom is this due to *extravagant field operations in game production*. . . . Usually it is due to the very high scale of living and entertainment on the preserves in question. Elaborate development of home site, beautiful landscaping and such activities should not be confused with field developments for game.[86]

Stoddard insisted that the management steps he took in the fields and forests of his southern hunting preserves, then, were perfectly compatible with management goals on government lands. And many of them did, indeed, translate to government land. But his technique and philosophy about environmental management arose within the context of a peopled landscape. Farming, firing, timbering, hunting, fishing—just plain living—were all incorporated, even integral, in Stoddard's land management. The landscapes with the most potential for plant and animal diversity, he thought, were those that locals had used for generations. Most such landscapes, of course, eventually came under the influence of an industrialized farming and forestry that lacked the restraint of the types of land uses Stoddard championed. Stoddard's ideal landscape, then, combined the small-scale agriculture of the people in the region with the protective ownership of the quail preserve set.

Although his techniques remained rooted in pre-industrial land use, Stoddard would eventually engage in newer markets to sustain the diversity he was after. The long Depression years caused many Red Hills landowners to curtail the extravagance and seek more substantial returns from their properties via modern agriculture and forestry. Again, they looked to Stoddard to guide them through the transition. His post–World War II engagement with the timber industry, and renewed interest in professional forestry, would bring his management full circle and force him to develop an even more thoroughly ecological philosophy of conservation.

CHAPTER 6

From Wildlife Management to Ecological Forestry

In late 1941 two very different storms came to bear on life in the Red Hills that would transform Herbert Stoddard's daily work and legacy. One was the worldwide upheaval of World War II, which created a strong, lucrative market for timber; the other, more circumscribed and quite literally a storm, rotated up from the Gulf of Mexico, leaving substantial wind damage in its wake as it passed through the Red Hills. The hurricane was relatively small, as such storms go. It was, however, a considerable disturbance event—to use the neutral terminology of ecologists—and it set into motion a major transformation in Stoddard's practice of land management. Stoddard was consulting in the Albany region when the storm hit on October 7. Once it passed, he made the normally two-hour drive home in ten hours, cutting fallen trees out of the road along the way. He finally arrived to see his house partially flattened by a yard oak, but his wife and son safe. It was only a matter of days before he considered the opportunities presented by the storm.

In his estimation, the forests of the Red Hills hunting preserves had long needed a good, strong wind. Increasing tree density was beginning to shade out valuable understory grasses, and wildlife struggled to find sources of food. On top of that, the landscape aesthetic that attracted so many to the region in the first place—open woods with a distant view—was no longer the norm. The blow-down from the hurricane would finally alert landowners to the advantages of harvesting timber. Stoddard noted to his friend Henry Beadel that, despite the hardship that accompanies such an occurrence, "after the fallen pine is removed the stands will probably be even better than before, as most of us were getting badly choked up with pine timber."[1] As counterintuitive as it may seem, the hurricane was just the sort of disturbance event that would help to restore the forests' declining aesthetic and ecological value.

Stoddard had been recommending that preserve owners thin their forests for the previous decade or so. In *The Bobwhite Quail*, he wrote, "As the trees mature they should be cut for lumber, so that the open, parklike nature of the woodland can be maintained."[2] In succeeding years he repeatedly urged members of the Cooperative Quail Study Association to open up their woodlands by harvesting timber. By 1937 he warned in his annual report that the "over-density of pine stands has become one of the most serious problems confronting owners of scores of game preserves in heavily forested sections like Thomasville, Georgia and Coastal South Carolina."[3] Stoddard came to think of the conservative harvesting of timber as a necessary component of effective longleaf woodland management, but he and his employers had seen the toll that loggers could take on the coastal plain landscape. The real problem was developing a method to select, cut, and remove individual trees from the forest without damaging its aesthetic and ecological value. But in a larger region moving toward an industrial model of forestry, a solution was not close at hand.

Following his work on fire in the longleaf pine region, Stoddard reengaged the forestry profession in the late 1930s and 1940s in an effort to integrate forestry and wildlife management into a single method of land management for the region. What he came up with most resembles what foresters today call "ecosystem management," or "ecological forestry," a method that values noneconomic resources by mimicking ecological disturbance patterns in harvesting timber and establishing tree regeneration.[4] Though he never placed such a label on his method, today's conservation ecologists recognize it as such.[5] In historical perspective, Stoddard's practice of forestry presents a stark contrast to the dominant industry-oriented silviculture of the post–World War II South. Stoddard's turn as a forester is important because it allowed him to refine his land management for biological resources and simultaneously engage industrial production. This was, again, a pragmatic move; second- and third-generation landowners were simply not as wealthy as their predecessors. If Stoddard wanted to continue in his capacity as an independent consultant he had to squeeze more income from the land for landowners. But at the same time, he saw the engagement of commercial timber markets as an opportunity to continue to shape his landscape, rather than a hindrance. And yet, his method of forest management came in direct conflict with the prevalent trends in southern forest management.

Several developments prior to World War II facilitated the spread of Stoddard's approach to land management outside of his Red Hills region. The

CQSA network, of course, was vital to the proliferation of his methods and ideas, and he continued to make inspection trips throughout the region before and after the war. In addition, Stoddard entered into a real estate partnership in 1936 that specialized in piecing together hunting preserves. His partners, Richard Tift of Albany, Georgia, and Jack Jenkins of Charleston, South Carolina, were already active in plantation real estate in their respective territories, and placing these three names together on one letterhead gave them a significant comparative advantage over other brokers seeking large landholdings in the Depression-era market. For the next two decades they became a powerful force in southern real estate, and one of the primary entities through which large, contiguous tracts of land came together into hunting preserves.

Tift and Jenkins handled the business end of the operation, and Stoddard put his feet to the ground, inspecting available lands for their game potential and drawing up land management plans for new owners. According to Stoddard, he was distinctly positioned for this sort of work because his work with the CQSA "has taken me all over the Southeast, so I know the districts best adapted to upland game quite well . . . and land examination previous to purchase has been one of my big branches anyhow for several years."[6] In addition, this real estate venture would finally give him a substantial income to plow back into his research. Stoddard explained to Aldo Leopold that while he was out on his inspection trips, "Komarek takes up the slack, and I am able to keep him on at a decent salary through putting back a portion of the real estate earnings into the treasury of the Association. Should have done this years ago."[7] These real estate contacts, along with the CQSA, composed a private network to facilitate the spread of his wildlife management methods, and they would soon help extend his method of forestry as well.

If the ecological disturbance of the 1941 hurricane could help enhance the aesthetic and ecological value of the forests, then the political disturbance of war could surely help to realize the forests' economic value. World War II spurred an interest in southern forests not seen since the cutover of 1880–1920. Demand for wood products—poles, pilings, and boxing materials were particularly desirable—far outstripped on-hand supply in the early years of the war, and the War Production Board, the Forest Service, and numerous forestry organizations worked closely to ramp up production. They set out to tap both public and private forest lands, and recognized the South's privately owned forests as a major source of wood. Government demand was so high that fears actually circulated about the condemnation of wood fiber on private land. That never happened, but the government did guarantee prices that would attract all but the least money conscious of landowners.[8]

Like the hurricane of 1941, the war presented an opportunity for Stoddard to learn more about silviculture and to further shape coastal plain forests to his liking. It also gave preserve owners an opportunity to enter a lucrative wartime timber market, while simultaneously deflecting growing local concerns about vast tracts of land sitting "idle" during a time of national emergency. Stoddard wrote to Jack Jenkins that "already, criticism is being voiced" that the preserves were not contributing to the war.[9] But many owners were, indeed, sensitive to local protests. He told another colleague that "most of their [wildlife] programs have been drastically curtailed and the owners are turning their attention largely to such things as farming and timbering as an aid to the war effort, and only doing enough for the game to maintain their present stocking."[10] Owners cut back on their wildlife work so much, in fact, that Stoddard and Ed Komarek disbanded the Cooperative Quail Study Association in 1943. Komarek undertook a "Farm and Game Service" to advise preserve owners on expanding agricultural production without detriment to wildlife, and Stoddard entered full time into the forestry and timber business. Neither had any qualms about leaving the CQSA behind; Komarek felt that "such large scale agricultural operations are very necessary" to the war effort, and Stoddard argued that both he and the owners "should be active with some essential work."[11] They were sincere about war production—and about making a living during tight times—but they also had ulterior motives for taking on the work themselves. Both Stoddard and Komarek knew what industrial agriculture and forestry might mean for the region's biological resources, and at least they could direct production with a lighter touch. On the forestry end of things, Stoddard actually hoped to enhance the region's woodlands, and to do so required that he gain control of every aspect of the timbering operation. From the early marking and sale of timber to the actual on-the-ground cutting of trees and postharvest cleanup, he was involved in every step of the process. The destructive cutting of old, he insisted, would have no part in the harvesting of hunting preserve timber.

The war was a symbolic turning point for southern forestry, and an important spark that ignited Stoddard's interest in the profession, but it was largely a four-year anomaly in the broader management trends of the region's forests. Since the reforestation efforts of the 1920s and 1930s, southern foresters had been working out how to best manage pine forests for wood production, and by 1940 they could only agree on one thing: despite the rapacious cutting of the previous generations, there was a lot of timber in the region. This consensus on timber volume resulted from the Forest Service's Southern Forest Survey, a major landmark for forest management and utilization in the South. The survey started in January 1931 as part of a national timber inventory authorized by the McNary-McSweeney Act of 1928; the coastal plain section was headquartered at the Southern Forest Experiment Station in New Orleans. Inman "Cap" Eldredge, already a legend in southern forestry circles, came from the Superior Pine Products Company in southeast Georgia to lead the survey team in early 1932. This was a major undertaking, to survey every corner of the region and measure wood resources. One participant recalled the grueling nature of the work, writing, "Don't make the mistake of deprecating the work in the South on the count of 'easy topography!' . . . [Some areas] were traversed literally on hands and knees."[12] The work paid off, however, from the southern boosters' perspective. Within just a few years, the survey began releasing figures that resonated throughout the forest products industries.

The South was blessed—or cursed, depending on one's perspective—with the reproductive capacity and growth rate of loblolly and slash pine. Despite the efforts of many foresters during these years to regenerate longleaf pine within its former range, it was an uphill battle in a time of fire exclusion. According to H. H. Chapman, the eminent forester at Yale and longtime leader of the Urania Experiment Station, "when established in competition with longleaf pine seedlings, loblolly pine will invariably suppress [longleaf] unless it is itself killed out by fires."[13] Major seedfalls every two or three years quickly seeded in many former longleaf lands, and loblollies, in particular, invaded cutover land in droves. Most southern foresters considered such a fertile, fast-growing pine preferable to the fickle, slow-growing longleaf. Philip Wakeley, an early researcher at the Southern Forest Experiment Station in 1924, did not "think that even those of us in research working on the natural reproduction problems realized the tremendous reproductive potential of the southern pines, longleaf excepted. It's easy compared to most parts of the world."[14] In addition to encouraging the natural reproduction of loblolly and slash, state forestry commissions established coastal plain tree nurseries and promoted the planting of the fast-growing seedlings. Loblolly and slash became so prevalent so quickly that most foresters simply gave up on longleaf pine.

This made sense in the short term, but the conversion of the coastal plain uplands to loblolly and slash pine presented serious complications for long-term forest management. They were both what today's forest ecologists call "off-site" species in the upland coastal plain. That is, they developed under different evolutionary circumstances from those of longleaf. The historic range of loblolly was primarily the southern piedmont, or the rare hill and hammock of the coastal plain that escaped frequent fire. Slash pine, on the other hand, was a solid member of the coastal plain community, but it grew mostly in lower elevations where seasonal water usually kept fire at bay. These areas might burn during severe drought, however, and both loblolly and slash adapted by growing a protective layer of bark in their first ten to fifteen years of growth. But neither species could withstand fire before that point, so their reproductive capacity depended on the absence of fire rather than its presence. By the late 1930s, then, foresters were developing management practices based on the behavior of two species that were reproductively unfit for the fire-prone upland coastal plain.[15]

With the survey results trickling in throughout the 1930s, foresters began to investigate which management practices would lead to the fastest growth and most abundant harvest of wood fiber.[16] When the war placed a greater demand on the region's supply, a management debate surfaced through forestry journals and bulletins that caused a brief but important controversy that would help determine the dominant forest structure in the coastal plain well into the future. The core questions were how to best harvest southern pine and how to encourage its regeneration, two problems that American foresters had long thought about in other forest systems. To no one's surprise, southern foresters imported the most prominent answers from other regions, but they took on a regional hue because of the unique ecology and land-use history of the southern coastal plain. Two management systems, in particular, came to dominate the debate, and they both had powerful proponents. H. H. Chapman supported what was known as "even-aged management," and the Forest Service's Southern Forest Experiment Station pushed "uneven-aged management." A slightly altered version of Chapman's system eventually carried the day, but that was not an inevitable outcome in the early 1940s.[17]

Both Chapman and the Forest Service hoped to systematize sustained yield forestry in the South through natural regeneration. They just couldn't agree on the details. For Forest Service researchers, uneven-aged management, which referred to the selection of individual trees for harvest, with the cutting spaced in ten- to twenty-year intervals, seemed to make the most sense. In a typical growth cycle of about sixty years, they argued, a landowner could cut three or four times and still have regenerative growth coming on to get a head start on a new cycle. Take, for instance, a piece of old-field land that seeded into a dense stand of loblolly pine over a period of five years. The oldest trees in that new stand would likely become dominant, and could be thinned after about twenty years. This first thinning would release the smaller trees, and in another ten years or so, you could take out the most marketable trees again. This second harvest would open up the stand even more, allowing sunlight to reach the forest floor, where new seedlings would regenerate and establish new growth underneath a canopy of maturing pines. When the oldest pines in the stand reached economic maturity, you could cut them out and begin the cycle again with the regeneration already captured. It all seemed to make sense, but this system was contingent on the exclusion of fire, a prohibition that would create all sorts of problems for a stand of multiaged loblolly pines.

For many reasons, H. H. Chapman considered selective cutting a "dangerous innovation" in these new loblolly pine forests. The problems hinged on the ominous-sounding "threat of hardwood invasion."[18] Absent periodic fire, hardwood species would almost inevitably sprout alongside loblolly, eventually shading out the pine and becoming dominant. But the use of fire in an uneven-aged stand of loblolly would kill any young reproduction. As Chapman put it, "normal, or ordinary winter fires, even at ten-year intervals, would largely defeat the system of single tree selection by wiping out the existing reproduction. But there is real danger that in the complete absence of fire the system would ultimately defeat itself" by allowing hardwoods to take over.[19] The uneven-aged management of loblolly pine presented a dilemma with no desirable outcome.

Chapman's way out of this dilemma was to manage for even-aged stands, using fire at prescribed intervals to control the hardwoods. Even-aged management was actually a form of clear-cutting, but the political baggage that accompanies that term today tends to obfuscate the nuances and logic of Chapman's research. It went something like this: on a piece of cutover land, Chapman recommended burning over the cleared patch to prepare the seedbed for loblolly regeneration, thus knocking back any emerging hardwood competition. Loblolly was a prolific producer of seeds, and with a few standing trees would quickly reseed the cutover. Where seed trees were not available, he recommended hand planting. After about ten to fifteen years of growth, the hardwood sprouts would likely return again, which required a light, winter-season ground fire to take care of them. By this time, a protective layer of bark would shield the loblolly from the fire. From this point until the final cut, Chapman recommended selective cutting every ten years, to be accompanied by another round of burning. In many ways, his system did not much differ from selective cutting, with the notable exception that a fire in ten-year intervals made loblolly reproduction impossible. The periodic selective cuttings culminated when the remaining trees reached sawtimber class, which in most cases was economic maturity. This final cut was the clear-cut stage, after which the process would start again. Ideally, landowners would practice this system on a large acreage, where they would have several stands at various points in the cycle.

In comparing the two systems, Chapman boiled the choice down to a matter of desirable forest conditions. A manager of southern pinelands "must make two decisions; first, whether he desires this transformation to hardwoods or wishes to prevent it, and second, what measures to take in either case to carry out his purpose."[20] The former was not really a decision at all—most hardwood species of the upland coastal plain were worth very little on the market. In regard to the latter, he argued that the only way to outflank the advancing hardwoods was even-aged management with periodic fire. And if a sustained yield of loblolly pine was the goal, he was largely right. This was the predicament of managing for a species with no reproductive tolerance for fire in the fire-adapted southern coastal plain. The fire history of the region made it nearly impossible to manage for loblolly without clear-cutting at some point.

Longleaf pine, on the other hand, had no such problems. But it did have trouble meeting market demand. Despite the expanding conversion of longleaf lands to "off-site" species like loblolly and slash, there remained a significant portion of the coastal plain in longleaf pine during the war years. And many foresters and industry officials had a great deal of interest in reforesting the range with its original inhabitant. Longleaf developed toward maturity at a slower pace than other southern pines, which gave it the strength and durability favored by lumber manufacturers. There would always be a market for longleaf no matter how slowly it grew, so forest researchers actually expended a great deal of effort to learn about the mechanics of reproduction, growth, and harvest.

W. G. Wahlenberg, a Forest Service scientist at the Southern Forest Experiment Station, synthesized all of the available research to produce the first major volume on the tree in 1946, entitled *Longleaf Pine: Its Use, Ecology, Regeneration, Protection, Growth, and Management*. While not quite a towering figure in southern forestry circles—like Cap Eldredge, Austin Cary, or Chapman—Wahlenberg was a meticulous researcher, writer, and field scientist. He spent years on the book, not only in the field, but also pulling together previously published research to present a full scientific and commercial portrait of the longleaf pine. The book would become the standard on all aspects of the longleaf pine for the next two decades. Stoddard reviewed drafts of the chapter on longleaf ecology, as well as several sections regarding fire, in early 1942, and with a few minor revisions thought the finished product "should clarify the whole matter and place the use of burning in its proper position in many respects." He told Aldo Leopold to be on the lookout for this "outstanding contribution to the knowledge of Southeastern ecology."[21] Indeed, this was the synthesis that longleaf admirers had been waiting for, and its explanation of the constituent parts of fire-adapted longleaf pine woodlands was the most ecologically sophisticated treatment available by far.

In terms of management recommendations, however—the part of the book that would make the largest impact on the land—Wahlenberg drew on market considerations as his primary guide. As Forest Service chief Henry Graves put it in the foreword, "The primary purpose of the treatise is to aid the timber grower . . . to secure reproduction and maintain a high standard of volume and quality growth, and thereby obtain maximum long-run financial values for timber growers."[22] Worthy goals, but with a tree like the longleaf pine, such considerations could only lead to frustration in a profession becoming more and more closely aligned with industrial efficiency.

The problem with longleaf forestry was regeneration and growth. Young seedlings were not very competitive without frequent fire, and even with it, they grew slowly. The planting of longleaf led researchers nowhere; they could get some seedlings to take hold, but, according to Wahlenberg, "it has so far been almost impossible to obtain early height growth or satisfactory survival in plantations . . . many plantations not only fail to begin height growth in their third to tenth years, but are understocked because of losses from brown spot [disease], vegetative competition, and other causes in the first decade after planting."[23] Attempts at efficient natural regeneration did not fare much better: "Regeneration of longleaf pine has been rarely accomplished. In fact, the reproduction of this species has been so irregular and uncertain, and its natural controls so imperfectly understood, that successes and failures have been difficult to explain."[24] The problem had little to do with their knowledge of longleaf regeneration; Wahlenberg went on to construct a highly refined discussion of how the tree reproduces and grows under natural conditions. The real problem was that longleaf pine was not suited to a market obligated to the fastest turnaround possible.

Longleaf pine is a plodder. It seeds in, takes root, and puts on growth at its own pace, and only under the right environmental conditions. The first requirement for successful regeneration is a large seed fall, which happens only every seven years or so. Longleaf seeds are a good food source for a host of forest animals, and a light seed fall would be quickly gobbled up by fox squirrels, birds, and rodents. When a seed fall does saturate the range during a heavy mast year, leaving enough to escape predation, seeds need bare mineral soil to germinate. Cones usually drop during late fall, so ideally a fire would blow through shortly before to expose the soil and allow the seed to sprout and take root. Once in the soil, the seedling begins to extend its taproot deep into the ground to find a water source. This taproot eventually lengthens to ten feet or more, burrowing for two to ten years before the seedling begins to put on growth above ground.

In the meantime, the exposed portion of the seedling sits in wait in the "grass stage." Looking like a coarse, green bunch grass to the untrained eye, the longleaf seedling grows its namesake needles of ten to fifteen inches within a year or so of germination, thus protecting the bud from fire. Once protected, the seedling can easily withstand fire until the taproot signals it is ready to send up new growth in the form of "candles," white growing tips that within a few months turn to brown, scaly bark. While in the candle stage, the tree is slightly more susceptible to intense fires, but that time is short. As Wahlenberg found, "in one test, tissue paper placed around the buds of seedlings 1 to 3 feet high was not even scorched though the needles were scarred to within 3 inches of the bud."[25] From the candle stage, the tree puts on growth at a fast clip. But the time and difficulty required in reaching that stage did not impress most foresters interested in fast, efficient, simplified reforestation. The primary frustration with longleaf pine by the 1940s, then, was not a lack of knowledge about the tree; it was its incompatibility with the need for control and simplification of forest resources under an increasing scale of industrial production. The longleaf pine was fast becoming an anachronism in the modern coastal plain.[26]

The complexities of longleaf regeneration and growth, along with the problem of market demand, ultimately led foresters to embrace the faster-growing species of loblolly and slash pine. Some observers recognized the potential management problems with growing off-site species in the coastal plain, but market incentives overshadowed any misgivings about turning their backs on the slower-growing longleaf. In fact, the expanding pulpwood industry rendered debates over the structure and management of coastal plain forests largely moot within a decade of war's end. On countless acres, the forestry methods of the pulp and paper industry replaced those geared toward the lumber industry. The trend was not new to the postwar period; kraft paper mills moved into the region in the 1920s and 1930s to produce commercial grades of paper. During that time, the U.S. Forest Products Laboratory in Madison, Wisconsin (where Aldo Leopold worked for a short time in the 1920s), as well as Georgia's Charles Herty, began refining chemical processes to produce newsprint and white paper from young southern pines.[27] The industry's move south sped up when the results of the Forest Survey alerted northern paper companies to the region's abundant wood resources. Cap Eldredge recalled the survey's influence on the pulp and paper industry, saying,

> The very time when [Charles Herty's] experiments had been done and he was broadcasting them to the world, we had completed enough of the forest inventory of the South, the first survey, to publish the results, and the results were astounding. No part of the country, not even the South—certainly not the South and perhaps not the Forest Service—had any idea of the amount of timber that we had here, how fast it was growing, and how universally it was distributed, and what state of development it had reached. When these two things came to the eyes of paper people all over the eastern part of the country . . . that started the movement South right away.[28]

Again, loblolly and slash pine were prolific colonizers when given the opportunity. And not only had the anti-fire campaigns been successful in many areas of the South; the absence of longleaf pine across the range meant the absence of its highly resinous needles to help carry a fire. The resulting stands of loblolly and slash, in the age classes ideal for papermaking, were highly desirable to the pulp and paper industry.

Paper mills began popping up along the Gulf and Atlantic coasts in the mid-1930s—Union Bag and Paper in Savannah, Georgia; International Paper in Panama City, Florida; St. Regis Paper in Pensacola, Florida; and St. Joe Paper in Port St. Joe, Florida, just to name a few. By 1955 there were twenty-seven paper mills in Georgia, Florida, and Alabama alone, and wherever a mill rose, the company was sure to purchase as much land as possible and implement their own version of forestry to feed the mill.[29] In 1956, for example, fourteen companies owned over 2.6 million acres of land in Georgia, most of which was located in the coastal plain.[30] From 1935 to 1955 pulpwood production increased from 2 million to 15 million cords regionwide, the price rose from $3.50 to $15 per cord, and production was nowhere near potential.[31] In just two decades, the pulp and paper industry came to dominate both the southern forest products industry and the land on which it relied.

Industry-owned land only provided a fraction of the necessary cordwood required to feed demand. In order to prosper, the pulp and paper companies had to rely on private landowners for raw materials—of the 3.5 million cords of pine pulpwood produced in Georgia in 1956, only 473,793 came from company land.[32] The vast majority of the region's forests still lay in the hands of small landowners, and in the years after the war, practicing foresters were motivated by one common purpose—as Florida's chief forester put it: "To reach particularly the small landowner and persuade him to handle his timber in such a manner that he will be able to produce continuous, profitable crops."[33] A number of state and private organizations jumped on board, and foresters became a fixture at local events throughout the region. The Southern Pulpwood Conservation Association (SPCA), based in Atlanta, was one of the most visible organizations. Formed in 1939 to protect and promote the industry's interests, it was soon covering the region with literature, producing and screening forestry films, sponsoring summer forestry camps, preparing fair exhibits, and hosting demonstration days on industry land.[34] The industry became enamored with the region's potential so quickly that they were playing catch-up to assist landowners in the newest methods of efficient forestry. Demand was such that, according to one company forester, "the problem in general is not to find a market for pine pulpwood, but to find the raw material to be marketed."[35] Companies held vast tracts of land, but to really reach productive potential, they would have to bring smaller landowners over to their brand of forestry.

Pulp and paper foresters subscribed to even-aged management, yet implemented some significant variations to Chapman's methods. They were leery of fire, and in the early years only used it to prep a site before planting. And when herbicides like 2,4-D became commercially available after the war, they eliminated the need for fire altogether. Pulp and paper foresters also managed on a short rotation, with few selective cuttings. That is, they planted trees, let them grow for thirty years, clear-cut them, then planted again. For their purposes, it became a remarkably efficient system. As one Union Bag forester explained it at the regional meeting of the Society of American Foresters in 1954, "since Union Bag's objective on their lands is to produce the maximum volume of chipable wood fiber per acre, and since we feel that the pine stands of the Southeast can best be perpetuated and managed in large even-aged stands, we intend to manage our timber in even-aged stands and on relatively short rotations."[36] The point was to erase the environmental uncertainties inherent to growing their raw material. Nonindustry foresters like Chapman were after the same thing, but they were interested in doing it within the confines of natural processes.

In contrast, figuring out the dynamics of fire and natural regeneration simply did not fit within industry goals. Pulp and paper's interest was in relocating the certainty of the factory floor to the forest. With that goal in mind, the Georgia Forestry Association in 1946 described in pamphlet form "What a Small forest should look like: Such a forest has no overripe trees, past their best growing years. . . . The litter and lack of sunlight on the floor have weakened and killed all grass and weeds. Nothing—neither grass nor young trees—grows on the floor of a closed forest in which trees are of the same age."[37] This was forest farming at its best, a biological factory for spitting out wood fiber.[38] Indeed, the Union Bag representative felt that "foresters working in the pine stands of the Southeast are in a unique position. We no longer need affiliate ourselves with the conservationists in order to justify our existence. Our activities are a money making segment of the business enterprises with which we are associated."[39] The purpose of industry foresters was to make money for their employers, and to do so required a singular devotion to their efficient methods of forestry.

Pulp and paper, however, was never without its critics. By the 1940s the industry was a constant irritation for conservationists like Stoddard. While on his real estate inspections, he became increasing aware of its transformative effect on environments across the South. As early as 1937 he informed Jack Jenkins and Richard Tift that southeast Georgia was largely lost to anyone interested in wildlife, writing, "Due to the activity in the paper business both at Brunswick and Fernandina, lands are rapidly being bought up in that district, so much so that . . . we should avoid wasting a lot of time in making detailed examination of available tracts, which might at any time be gobbled up by the paper interests."[40] Indeed, the pulp and paper companies were Stoddard, Jenkins, and Tift's major competition for large tracts of land throughout the coastal plain. Time and again, Stoddard informed prospective buyers of areas that had "gone all out for forestry," and of "more lands becoming unsuited to quail over . . . the whole South."[41] By 1947 even the head of the U.S. Fish and Wildlife Service could not find a suitable tract of land for his brother. In response to an inquiry from Ira Gabrielson, Stoddard wrote that he knew of a suitable place two months prior, but "this property was sold to a paper company, and I personally feel that the fine stock of wild turkeys and quail will not long survive."[42] Stoddard's exasperation with pulp and paper came partly from an inability to do business—they were removing good tracts of coastal plain land from the hunting preserve market. But more importantly, he knew these lands would quickly transition into loblolly or slash pine plantations where understory biology suffered. For Stoddard, his barometer of land health centered on wildlife and their requirements for food and shelter, and he considered the forests of the pulp and paper industry as sickly at best.

Members of the southern hunting preserve community were not the only voices that opposed pulp and paper. The forest industry sectors that utilized mature timber also felt threatened. One member of the Georgia Forestry Association worried as early as 1937 that "the newsprint and white paper mills will soon come and it is easy to understand the anxiety of the naval stores operators, sawmills, veneer and pole producers, as well as other consumers, as to their timber supply."[43] There was a real fear that pulp and paper would take over the southern forests, forcing landowners to convert to thirty-year rotations, and thus eliminating the supply of sawtimber. In some localities, the lumber interests actually appealed directly to landowners. In Troup County, Georgia, for instance, local sawmills ran a series of full-page advertisements in the local paper in the spring of 1944, counseling landowners that "a word to the wise is sufficient. You wouldn't sell all your Pigs before they become Hogs. So why sell all your Pine Saplings Before They Become Timber?" A week later, they became more forceful: "Listen! We told you how we feel about letting the Pulpwood folks Clean our County of all its Sound Trees. We Have an 'Axe to Grind.' We want to continue to stay here . . . NOW! Troup County Lumber Industry."[44] Without landowners growing timber on a long rotation, local sawmills had few places to turn.

Other criticism came from within the forestry profession. Pulp and paper's simplification of the region's forestland was a real concern for those foresters who had spent their lives trying to understand how the forests worked, no matter how corrupted it had become. H. H. Chapman became especially disillusioned with the direction of southern forestry. Though his research was almost always in service of the forest products industries, Chapman considered pulp and paper forestry mind-numbing in its myopia. In a 1954 *American Forests* article where he asked "Do We Want a 'Pulpwood' Economy for our Southern National Forests?" he upbraided the industry for thinking only of the their own interests:

> Many, if not the majority, of the large southern corporations, faced with the problem of supplying their huge investments in plants with adequate raw material, are tending towards the management of their own lands on short rotations up to 30 years, cutting the crop clean, and repeating the process by planting or from seed trees. . . . But the comparative values of the crops by each method are heavily, lopsidedly weighted in favor of the method of thinning for pulpwood and producing as an end product a crop of high quality sawlogs. . . . But when more pulpwood is the dominant urge, the drive is to produce the greatest yield of this single product from which the mills derive their profit. Carried to its logical conclusion, this policy would neglect all other economic and public interests in order to achieve this one goal.[45]

The general fear in forestry and timber industry circles was that landowners, both public and private, would move exclusively to producing wood fiber for pulp and paper, leaving other industries in the lurch.[46]

Such was the fluidity of southern forest management during the years surrounding World War II. There was a common assumption that ran through all of these management regimes, however. Foresters from academia, the Forest Service, and pulp and paper were all attempting to manipulate coastal plain ecology to produce a steady stream of raw materials. Some, like Chapman and the Forest Service, attempted to work with and learn from the region's evolutionary ecology to meet their goals, while others, like the pulp and paper industry, simply imposed an industrial logic onto the landscape. As a result, the management systems of the former left some room for wildness as a by-product of production, while that of the latter left very little. But they all operated under a model of industrial production.

The system that Stoddard developed in the Red Hills and spread to other choice areas, on the other hand, managed specifically *for* wildness, which made it unique in the longleaf woodlands of the southern coastal plain. As will become clear, he borrowed many of the methods of professional foresters and tapped into industrial markets, but he operated outside of institutional management debates and had a different set of desired outcomes for the forests he managed. The system he developed and practiced in the

1930s, 1940s, and 1950s, in essence, left room for production as a by-product of wildness, rather than the other way around.

Stoddard had two distinct advantages over most practicing foresters when he entered into forest management: he had no imperative to maximize timber production, and he did not have to start from scratch. The turn-of-the-century timber boom had largely bypassed the region, leaving many stands of pure longleaf pine with its grassland understory intact. Instead of learning about the dynamics of a replacement forest—the task of most southern foresters—Stoddard could, in many cases, work with a forest composition that had been in place for thousands of years. Some of the woodlands he worked over were multiaged old-growth woodlands, the oldest longleaf trees being anywhere from 300 to 450 years old; other woodlands might be former turpentine lands where natural reproduction was slowly replacing worked-out or dead "face trees"; still other stands were second growth of either pure longleaf or a longleaf-loblolly-slash mix. Past land use usually determined the forest structure, and the Red Hills had evidence of most all of the older southern land uses—except that of the industrial cut and the ensuing period of extended fire suppression. This ecological continuity was critical as Stoddard began to think about how to harvest and regenerate timber. Most importantly, the maintenance and restoration of longleaf woodland was far less complicated with an uninterrupted fire history.

Though he only began to directly manage forests in the 1940s, Stoddard had long commented on forest resources in relation to wildlife management. One of his consistent mantras from the beginning was that no system of land management could expect to produce the highest amount of more than one species, whether it was wildlife or wood. If a landowner wanted the most out of timber production, he or she would likely diminish many wildlife populations. On the other hand, those seeking the upper limits of wildlife numbers could not expect a full stocking of timber. As early as *The Bobwhite Quail*, he wrote that in regard to quail, "it is obvious that nothing like a *maximum* lumber crop and *maximum* quail crop can be produced on the same ground."[47] Though he rarely leveled public judgments on market-oriented foresters—they had a job to do and Stoddard usually respected that—he did urge resource managers to use a little restraint. As he did with agriculturalists, he hoped the region's foresters would work to integrate wildlife and forest resources.

This concept of integration, of not emphasizing one resource to the detriment of another, became a central tenet of wildlife management in the

1930s. Many practitioners spoke of the idea, but, as with so many management concepts, Aldo Leopold thought it through with the most care. Leopold's career, of course, began as a forester with the Forest Service, and the management of forest resources was perhaps most dear to him. His most famous treatment of what foresters would eventually call "integrated management" was a two-part article in the *Journal of Forestry* on the *Dauerwald* forest movement in Germany. *Dauerwald* literally means "permanent forest," and was a movement to turn back centuries of single-species production in the German forestlands that had left their soils, wildlife, and timber resources in an exhausted state. Leopold's 1935 journey to Germany to study *Dauerwald* and the relationship of game management to forestry produced his most definitive statements on the subject to date.[48]

Leopold was impressed with the German efforts to cultivate a mixed forest, but the continued separation of forest and wildlife management made him realize the importance of merging the two activities. The Germans had a longtime devotion to spruce and deer—a devotion that created awful conditions for most everything else in nature. The even-aged management of spruce shaded out the understory, leaving deer to forage on every available leafy material, stripping the forest clean. As one German forester expressed it to Leopold, their "love affair" with deer led them to rectify this situation by setting up feeding stations, thus sustaining a deer population that the forests could not support. The *Dauerwald* movement attempted to fix the problems in the forest, but it also necessitated fencing deer out of the forestlands so they would not forage on new vegetation. As a result, Germans still fed the deer herds, and problems of overpopulation persisted. As Leopold's biographer, Curt Meine, put it, "The deer found themselves caught literally between one fence that excluded them from the new 'permanent forests' and another that excluded them from cropland, doomed to walk to their feeding stations on soil-sick forest floors."[49]

Leopold had his experience with deer irruptions on the Kaibab Plateau to draw on, and the situation in Germany confirmed for him that the fields of forestry and game management needed a thorough integration. His article on *Dauerwald* opened with a revelation: "The observer is soon forced to the conclusion that better silviculture is possible only with a radical reform in game management. Later, as he learns to decipher what silviculture has done to the deer range, he also grasps the converse conclusion that better game management is possible only with a radical change in silviculture. Germany, in short, presents a plain case of mutual interference between game and forestry." This situation was not unfamiliar in the United States. No longer could resource managers make the "uncritical assumption, dying but not yet dead in America, that the practice of forestry in and of itself, regardless of what land or how much, promotes the welfare of wildlife."[50] Leopold's observations in Germany had a dramatic influence over his thinking about American environments and their use. As he shared with an audience after his return to Madison, "All land-uses and land-users are interdependent, and the forces which connect them follow channels still largely unknown."[51]

This type of interdependent management just seemed like common sense to anyone who valued wildness in an age of industrial production. And when Stoddard began cutting trees and selling them on the market, he sought, rather innocently, to uncover those unknown channels that connected various land uses and land users. Stoddard best summed up his land management philosophy in a pithy parenthetical comment to Leopold in 1946. Of he and Ed Komarek's postwar work, he wrote, "We are working hard (timbering and farming more than wildlife work, but they are all woven together anyhow.)"[52] Indeed, as Stoddard, Leopold, and Komarek saw it, these land uses were all of the same natural fabric, one never cut off from the other, and with the aggregate productive landscape retaining the promise of wild nature.

As for his system of forest management, it was complex in the abstract, but in practice everything centered on two simple components: the frequent use of fire and the conservative marking of timber. In a typical multiaged stand of longleaf, for instance, he would study each candidate for harvest, paying particular attention to the condition of the tree and the type of canopy opening that would result. The major purpose of his forestry work, besides creating income for landowners, was to bring sunlight to the grassy understory, thus providing food for wildlife and maintaining a consistent fuel load to successfully carry a fire. A canopy opening too large might not gather the necessary needle fall to carry a fire, and thus allow hardwoods to sprout; an opening too small, on the other hand, might shade the grasses so important to wildlife. A key variable that helped determine the amount of timber to take from the forest was the amount of timber available in the forest, an obvious point perhaps, but important nonetheless. Stoddard began his marking system at a time when the Red Hills forests were loaded with timber, and he was able to mark a significant number of saw logs and poles for market without significantly diminishing the total volume.[53]

After creating openings in the forest, Stoddard turned his attention to longleaf regeneration. Again, fire was central to this task. "Fortunately, the composition of a pine forest in the making," Stoddard advised, "can often be controlled by the time and frequency with which fire is used, so it is well to use fire in a way that will give the longleaf an advantage in the seedling stage."[54] Following H. H. Chapman's findings at Urania, Stoddard advised "burn[ing] the area in late winter, as usual, until the year of a heavy fall of seed or 'mast,' so that the seed will have a good chance to reach mineral soil and germinate properly. As the longleaf pine produces a maximum seed crop at average intervals of about seven years, with one or two lesser crops between, it will be necessary to observe the developing cones closely so the time of a heavy production will be known."[55] Once recognizing a heavy seed fall, he advised burning the area from September 1 through early October to expose the soil. Some foresters recommended burning earlier in the summer, but Stoddard was concerned about the fate of ground-nesting birds and small mammal reproduction during a summer fire.[56] After capturing a heavy mast, he protected the area from fire until the seedlings developed their fire-resistant crowns about a year or two later. At that point, he resumed burning to protect the new growth from any encroaching rough. The result was a fire-maintained forested grassland with longleaf growth in every possible age and size class. In contrast to loblolly and slash pine, then, the fire-adapted longleaf permitted the practice of continuous uneven-aged management.

Being in the woods every day marking trees thrilled Stoddard to no end. In wartime correspondence he reiterated again and again how pleased he was with the forestry operations. Writing to Leopold in 1942, he declared that "some of our woods look wonderfully fine after being 'culled,' and I am doing personally what I have preached for years."[57] He explained a few months later that "all of our work is on a careful selective basis, and carried on with the welfare of wildlife and the future good of the forest in mind, and much of it consists of culling operations on worked out turpentined pine, crooked, suppressed and diseased material." Rather than high-grading a stand, as most selective-cut foresters advised, he took the weakest trees first. This opened up the stand, and as regeneration came in during subsequent years, he could take more of the high-grade timber. Stoddard's involvement from front end to back end was crucial in his estimation, largely because he did not trust anyone else to carry out the work with care. He continued: "I am literally up to my neck getting timber out . . . I personally mark and supervise the cutting of some ten million feet of pine lumber, several thousand poles and piling, crossties and perhaps five thousand units or more of pulpwood per year. . . . On the theory that a fellow can best instruct others if he knows the game himself, I do not regret having to do all the field work myself."[58] Stoddard's hovering presence did not make a logger's job easy, but the value of his timber made it worth the hassle.

The cleanliness of the logging job, in fact, was almost as important as the marking of trees. Stoddard put together a standard logging contract that must have looked alien to most loggers in the South. Prior to presenting a contract, he advised timber buyers in memo form that "the requirements governing the cutting of timber are so different on the properties in question from ordinary commercial timbering operations" that loggers should be prepared to abide by strict guidelines in cutting and hauling timber from the hunting preserves.[59] There were five basic conditions required of the logger. First, "trees will be felled with due regard to minimizing damage to trees left standing," and all limbs and tops were to be cut off in the woods and pulled back at least six feet from the base of living trees. This allowed the next fire to burn away refuse without damaging living trees. The remaining conditions were largely for aesthetic reasons. The logger had to cut the trees so that stumps were no higher than six inches, they could not use mechanical skidders to pull logs through the woods, they had to use existing roads to remove timber, and they were not to leave any litter or waste in the woods.[60] Seems reasonable enough, but this was not the type of supervision loggers were accustomed to in the woods. The goal was to maintain a consistent aesthetic experience before and after the harvest, so that an unsuspecting passerby would never know trees had been cut. If the forest looked the same as before a logging operation, Stoddard's reasoning went, then it most likely functioned the same.

While loggers may not have been overjoyed by Stoddard's presence in the woods, others were. Leopold was keen on Stoddard's move into forestry, and perhaps a bit envious. He had long known that Stoddard was in a unique position as an independent researcher and conservation practitioner, and this new phase in Stoddard's work was just the sort of integrated management he advocated. Leopold wrote, "Your metamorphosis to a practicing forester pleases and interests me. . . . I suppose you realize how rare an animal you are, i.e. how rarely any forester or game manager has practiced conservation without a subsidy."[61] Indeed, Stoddard was a rarity. While his colleagues in the biological sciences and natural resource management worked in bureaucratic lines of command, and were beholden to an often detached administrative order, he was able to carry out his work with very little oversight. This ability to practice science and conservation in relative isolation from academic, governmental, or industrial oversight largely accounted for the form his management took. The transition into forestry made Leopold all the more curious about the work of his independent friend: "I only wish I could tag along with you for a week or so, just to absorb the flavor. I'd just as soon do that as hunt turkeys with you, and that is saying a good deal."[62]

Leopold had already tagged along with Stoddard in the woods many times, and he thought any coastal plain landowner could learn from Stoddard's land management regardless of their goals. From his visits to the Red Hills, Leopold suspected Stoddard was up to something of a groundbreaking nature. When Daniel Hebard, a Philadelphia lumberman who owned a hunting preserve in southeast Georgia (as well as most of the Okefenokee Swamp), asked for an estimation of Stoddard's methods, Leopold was emphatic in his praise:

> I have just spent several days with Stoddard and came away with a conviction that he has been too modest about the conservation methods he has worked out for the Southeast. They are commonly regarded as applicable only to game preserves, but in my opinion he has developed principles which are equally applicable to lumber company holdings, national forests, and all other owners of coastal plain longleaf. . . . I of course am biased, for he is one of my closest personal friends. I am lecturing to my students Monday on the Stoddard method of handling Southeastern pine lands.[63]

This was an impressive endorsement, and Leopold recognized what few others did in Stoddard's landscape: it was the result of a broad-based management regime that combined the applied sciences of wildlife management, agriculture, and forestry, and could work for any landowner interested in conservation.

Others did not share Leopold's conviction. Most southern foresters were completely caught up in the throes of modern forestry. The year before Leopold's 1939 visit, the Southeastern Section of the Society of American Foresters came to Thomasville for their annual meeting, and most of them frowned on what they saw in the surrounding countryside. Stoddard and Ed Komarek were most hospitable, leading a group of sixty-nine foresters on a field trip through several of their finest longleaf woodlands. There was most likely audible grumbling in the woods—old-growth timber still on the stump, stands that were not fully stocked, with little patches of regeneration here and there. This was the antithesis of good forestry in the coastal plain, they likely thought.

Figure 11. Herbert Stoddard in the "Big Woods" of Greenwood Plantation, 1964. As the southeastern forestry profession increasingly turned to the clear-cut and plant model of forestry after World War II, Stoddard developed his own model for harvesting timber that took ecological considerations into account. Courtesy of Tall Timbers Research Station.

Frank Heyward, a forester for the Georgia Department of Natural Resources and an acquaintance of Stoddard's, wrote after the meeting to convey the general feeling of disappointment. "There seems to be an apparent misunderstanding on the part of a number of the foresters," wrote Heyward, "who apparently felt that they were to be shown various examples of good forest practices." Resurrecting the conflict between forestry and game management that both Leopold and Stoddard were attempting to reconcile, Heyward continued: "Those of us familiar with the ecology of our pine lands realize clearly that the most desirable forestry practices are incompatible with the most desirable forms of game management or vice versa, as was the case on the properties which we saw."[64] Indeed, Stoddard could agree with that, but he did not apologize for trying to do a little bit of both. He replied: "Certainly all we have ever claimed is that a great deal of timber can be produced right along with game management. . . . The [Red Hills] region speaks for itself as to that. I think foresters should be gratified that the two land uses dovetail as well as they do." Though this exchange took place before Stoddard actually started marking timber, he was already forming the basis for his method: maintain a multiaged forest with plenty of mature timber and a diverse understory, and never discount aesthetics. He concluded: "The best we can do is get a good stand of pine started and then handle the ground growth for the game. And in the pine stands conditions are more favorable for the game and hunting in the latter portion of the cycle than in the first part. Hence the aim of the game preserve owner should be to produce a good crop of mature high grade timber."[65] This was a recognition of how different economic pressures could produce different landscapes, but it was also a clearly coded jab at the recent proliferation of the short-rotation, even-aged management of pulp and paper.

Coming out of the Society of American Foresters meeting, it seemed that Stoddard was gearing up to take on the forestry establishment once again, at least in regard to the lands where his form of management was in place. He revised his standard article on fire in 1939 so that it directly addressed forestry practice, whereas before it only spoke to the use of fire on game lands. He wrote of regenerating longleaf pine through selective cutting and well-timed fire, and of maintaining or restoring "a forest of unexcelled beauty and quality, strongly influenced by the policy of human inhabitants, [which] persisted for centuries until destroyed by the lumbering and turpentine industries."[66] This type of rhetoric was not what the pulp and paper industry wanted to see in print. Not only did he promote a nonindustrial forest structure, but he also gave rural southerners credit for creating it. In making his specific forestry recommendations, Stoddard cited Chapman and the Forest Service, but even still, industry foresters feared game management might invade a public discussion on forest resources they were working hard to direct.

Frank Heyward—by this time the director of the Southern Pulpwood Conservation Association—again let Stoddard know of his increasingly irritated constituency. Of the revised fire article, Heyward wrote, "Several foresters whom I heard discussing this publication felt that you had somewhat overstepped the bounds of your professional field by mentioning so many points pertaining to forestry. They also felt that your recommendations of controlled burning were more liberal than they have been . . . and that such recommendations would tend to defeat their own efforts in organizing forest fire protection."[67] Despite a growing recognition of some beneficial uses of fire, remarks such as Stoddard's seemed dangerous to an industry continuing its long attempt to establish administrative control of the region's private forests. Their core tenet in this effort remained fire protection, and Stoddard's was a voice of opposition that carried with it a great deal of authority.

Heyward also felt uneasy about Stoddard's fondness for old growth. The concluding references to the region's "original" forests were, in Heyward's estimation, "dangerous in that it makes the reader forget this distinction" between controlled and uncontrolled fire. He continued, writing, "You must remember that ordinary uncontrolled fires seriously retard growth, and that although a longleaf pine forest would probably tolerate a number of ordinary fires, the age required for this forest to attain merchantable size would be greatly increased as a result of the fires. Of course no one cared how long a period was required for the development of our virgin forests."[68] Talk of maintaining and restoring stands of old longleaf was practically unknown to the southern forestry profession by this time. Heyward himself respected Stoddard's efforts in the forest, but as the new director of the SPCA, he had an industrial regime to defend.[69]

By this time, Stoddard simply did not care what foresters thought of his crossover into their profession. As was his way, he responded to Heyward with a long, detailed letter explaining his view of forestry and why he no longer felt tentative in critiquing the profession. "I have heard that several foresters feel that we have over-stepped the bounds of our field by commenting so much on forestry," Stoddard wrote, but "this has not bothered me in the least, due to the fact that foresters for many years have written and said a great deal about the management of game. In fact, after the precedent they have established we feel no hesitancy at all in saying as much as we like about forestry management."[70] It was a biting comment, really, and one slightly out of character. But Stoddard was confident in his management system, and he would not apologize that his management followed the guidelines set out by nature instead of industry. In defending his preference for mature forests, he slyly, and rather sarcastically, slipped into the rhetoric of a forester, writing of the higher prices he received for old, slowly grown trees. "If we only knew how to get this slow growth by the use of fire," he pondered, "I think you would make a special point of using fire in a way that would secure it." But regrettably, "trees on burned land may grow as rapidly as those on unburned," so he had to settle for the best quality timber he could get.[71]

While Stoddard guided untold numbers of foresters through the Red Hills forestlands after World War II, his workload during the late 1940s and early 1950s was much larger than he would have liked. As a result, he only made a few attempts to present his comprehensive land management system to a broader audience during those years. He presented a paper on the integrated management of wildlife, forestry, and agriculture at the 1956 meeting of the Society of American Foresters, and another the next year on forestry and fire at a Louisiana State University conference entitled "Special Problems in Southern Forest Management." In both of these presentations, he turned his attention to small landowners in the South. In his thirty years in the Red Hills, he had seen his hunting preserves become biological islands in a region under the spell of industrialized forestry and agriculture. Both of these presentations struck a sad yet undefeated tone, and they represent two rare occasions when Stoddard argued for more state support of private landowners, especially for those interested in increasing wildlife populations while maintaining productive farms and forests. He made clear that combining all land uses was possible—he had been doing it for quite some time. But the available advice from foresters, agriculturalists, and wildlife professionals "has often been conflicting, which makes for confusion" among farmers. He lamented that "reaching owners of small land units, and actively assisting with simple and effective combined programs of forest, wildlife, and farming, has never been accomplished on a large scale in the region."[72] In his thirty years in the Red Hills, the specialization of the natural resource professions had intensified to the point that landscapes that once harbored a diversity of land uses were ever more cordoned off for discrete purposes. As a result, formerly wild farms and woodlots were becoming practically barren of diverse flora and fauna.

Stoddard was particularly disappointed in the modern composition of southern forests. In a stinging indictment of the pulp and paper industry, he told his audience in Baton Rouge:

> The greatest future threat we see to forest game and wildlife is no longer wildfire. It is rather short rotation forestry, either of pines artificially planted or naturally reseeded. When such forests are of vast extent, and in full stands of even-aged pine as is usually the case of forests handled primarily for wood pulp, they are 'biological deserts' that can be inhabited by few creatures and no game to speak of. This is true whether they are control burned, or fire is totally excluded as is so frequently the case. Little food producing herbaceous vegetation, or even shrubs can grow on the heavily shaded forest floor, nor is there enough surface fertility or water in any case. There is little cover to shelter game, and very few insects. Unfortunately for wildlife the acreage in such forests is increasing at a rapid rate as more and more paper mills move into the Region.[73]

He was not unique in his condemnation of modern forestry, but his opinion by this time held a certain amount of moral authority. He was not only recognizing the value of biological diversity; he was also advocating for its preservation in the face of decline. Stoddard had witnessed a major transformation of the coastal plain landscape since moving to the South, and throughout the last two decades of his life became increasingly disheartened about the prospects for a wild future. He did, however, believe in the perseverance of nature. In one of his most biocentric moments, he wrote to an old friend in 1950, saying, "You would not know Georgia; it has truly had an 'agricultural face lifting.' Maybe some of the changes are for the better; but for this one there is too much emphasis on the genus HOMO and too little consideration given to the other inhabitants of the Earth. But old Mother Nature will correct this when she feels that it has gone far enough. I am not going to let brooding on such take the pleasure out of life as do so many naturalists."[74] All he could do was care for what he had control over and hope that a hint of wild nature would remain on the rest, biding its time.

Stoddard's method of forestry would not receive much attention until long after his death, but it persisted locally in the Red Hills and Dougherty Plain regions for many years to come. Leon Neel, a Thomasville local with a forestry degree from the University of Georgia, began an apprenticeship with

Figure 12. Leon Neel and Herbert Stoddard, ca. 1960. Neel began working for Stoddard in the forestry business in 1950 and became a partner soon thereafter. Together, they crafted a model of ecological forestry that has since become known as the Stoddard-Neel Approach. Courtesy of Leon and Julie Neel.

Stoddard in 1950, and he continues to practice what has become known as the Stoddard-Neel approach to forestry.[75] The South at large, on the other hand, continued down the path of industrial forestry as devised by the pulp and paper industry. Even the long-rotation forestry of H. H. Chapman and the Forest Service became contained to only a few pockets of National Forest land and timber company holdings. Pulp and paper had successfully organized their systems of tree management and procurement to such wide extent that a multiaged forest more than thirty years old became a curiosity to most folks in the rural South.

But the woodlands that Herbert Stoddard shaped and Leon Neel maintained became so much more than mere curiosities. They became powerful representations, much like Stoddard's old museum displays, of what nature had to offer. As environmental artifacts of Stoddard's long dialog with nature, they also represent the artful implementation of knowledge. As an aging Stoddard told one group interested in forestry and fire,

> Ever since I was interested in burning we had to make that decision on the basis of the land you were standing on. You can't possibly give a formula for burning on land you haven't seen. You have to go and visit a piece of land and see what it requires. You have got to apply it just as you would in cutting timber. You can't say you are going to cut ten per cent or twenty per cent of a hundred thousand acres. You've got to go and examine that land and somebody that knows what he is doing has got to do the work. And that's where the art comes in. Science is never going to solve that. This would be the art.[76]

The art came from living in the environment, learning how it works, and becoming a part of the processes that shape it, Stoddard's lifelong project. The art of management, to be sure, bears the value-laden mark of human design. Just as many foresters created a model of production based on ecological simplification, Stoddard represented his values in one based on ecological diversity.

CHAPTER 7

Bringing Agrarian Science to the Public

By the 1950s Herbert Stoddard had become something of a cult figure among American naturalists and scientists. Though less engaged in professional matters after World War II, he continued to host a steady stream of visitors of all stripes at Sherwood: academic biologists eager to see some of the finest remaining longleaf woodlands in the coastal plain; government officials usually seeking out subtle ways to subvert official management policy; wildlife managers looking for the latest field techniques; and, as always, ornithologists hunting down information on the latest bird migration or simply looking for good bird habitat. Stoddard, in other words, rarely performed his daily duties alone. With such widespread interest in his work and habitat, he came to recognize the potential to use the Red Hills as a public, teachable place, a biological reserve in which to learn not only about the longleaf pine woodland environment, but to learn how to mesh human choices with natural processes.

The idea for a formal research station had percolated in Stoddard's mind since the Cooperative Quail Investigation. The public controversy over fire, in particular, called attention to the confusion among so-called experts about the workings of coastal plain ecology. Stoddard found inspiration in the controversy in September 1931, while in the midst of an extensive correspondence with the American Forestry Association's Ovid Butler. After hauling Butler over the coals for his Dixie Crusader campaign, among other things, Stoddard ended one of many letters on a constructive note. He encouraged Butler and the AFA to "get behind my suggestion of an Experiment Station devoted entirely to a thoroughly scientific study of the effect of fire on plant and animal life in the Southeast."[1] Such a research station, in Stoddard's estimation, would operate independently from a large constituency with disparate interests, and would include scientists from a number of disciplines—botany, zoology, forestry, geology, chemistry, and more. Butler responded with measured enthusiasm, and pushed Stoddard to elaborate. "I am *not* interested," Stoddard replied, "in such research carried on by any agency that has been in the anti-fire propaganda business for a long period and has a vast structure of published and expressed opinion to uphold." The U.S. Forest Service, in other words, should stay far away. He continued, "I am not keen for the sort of so-called research that is organized to determine 'the *damage* that fire does,' for we have already had too much of that. I want to see research to determine the *effect* of fire. Such an investigation," he concluded, "should be a study in pure ecology . . . [which] might best be carried on under some great Foundation in which all the investigators might be entirely unbiased and interested in only the truth."[2] Like many resource specialists of the day, Stoddard sought to use scientific expertise to solve a complex socio-environmental problem, yet at the same time wanted to strip that expertise from its traditional arbiters—career government and industry professionals.

It would take over thirty years for Stoddard to bring his vision for a proper research station to life. But he finally did so in early 1958, with the establishment of Tall Timbers Research Station. Tall Timbers began as an independently funded center set up with the ideal that scientific knowledge, undistorted by political economy, would prevail on the decisions of public land policy makers and conscientious private landowners. Like Stoddard's career, it represents a bridge between pre– and post–World War II conservation science, as well as between the management of public and private landed resources. Rooted in the management ethos of the interwar years, Tall Timbers sought to apply the lessons of those decades to the complex questions of postwar resource management.

By the time Tall Timbers Research Station came to be, the American countryside was vastly different than when Stoddard made his way south in 1924. Changing methods of production were particularly alarming. Diesel-powered machinery and synthetic chemicals gave Americans unprecedented control over nature, and when coupled with the industrial organization of agriculture and forestry, led to sweeping landscape transformations. In addition, rural life was not as important in the national consciousness as it once was. The rhythms of urban life began to define the nation after World War II, and suburbia had already begun its steady march across the American countryside. At the same time, the nation was in the midst of a revolution in environmental thinking. The postwar environmental movement came from several places: consumers, worried about losing the newfound comforts of affluence, began to protest the environmental abuses of industry; suburbanites decried the disappearance of the remnant rural spaces they sought in their flight from cities; and the science of ecology expanded in popularity and had people thinking evermore in terms of connectivity between human action and environmental consequences.[3] Out of this physical and ideological transformation, the modern environmental movement emerged in full flower.

These national trends cannot fully explain the motivations behind Tall Timbers. Its founders were certainly influenced by the changing patterns of postwar production and consumption, but their approach to conservation remained firmly rooted in the agrarian science they had practiced for decades, a type of science that was infused with an agricultural ethic and followed the contours of rural life. The environmental movement can, however, help to explain Tall Timbers' broader significance. In short, professional resource managers struggled to manage for a broad range of new environmental values. Managers added concerns such as recreational access, scenery, and environmental health to a list of management goals that already included the traditional economic demands placed on landed resources. The modern environmental movement, in other words, changed the way resource managers approached land management, challenging the capacities of public agencies to keep up. Those responsible for keeping up were the primary audience for Tall Timbers during its early years. The new research station, after all, represented over three decades of accumulated knowledge about managing land for multiple uses. The landscape of the Red Hills had long been a manifestation of a certain set of environmental values—aesthetics, recreation, economy, ecology—and Stoddard and Komarek's postwar forays into forestry and agriculture only intensified their efforts toward finding an integrated system of land management. The modern environmental movement demanded management for all of these values, and Tall Timbers became an important conduit for experiential knowledge.

Fire was a particularly important subject of interest at Tall Timbers. Coastal plain ecology had changed so much in the first half of the twentieth century that only a few old-timers such as Stoddard and his collaborators could demonstrate fire's effectiveness with any confidence. It was not only a necessary ecological component in the coastal plain; it was also an ideal management tool. When knowledgeably used, fire could accommodate all of the environmental values of the postwar world, and there was no better example of its efficacy than the Red Hills. Skepticism of fire, however, remained the prevalent sentiment among both forest organizations and the general public, and fire advocates were "constantly put on the defensive" throughout the 1950s, according to Komarek.[4] Stoddard and Komarek had witnessed the use of fire shift from being a common occurrence directed largely by local rural dwellers, to a subject for specialists only, most of whom had little experience actually using it. They held on to a folk knowledge of fire that was indispensable for its continued use, and Tall Timbers became a critical reserve for that knowledge. This would become all the more important as a growing tolerance of fire, and even an enthusiasm for its use, would soon be a hallmark of ecological management. The small cohort of scientists and land managers in this corner of the South, then, not only maintained some of the most spectacular landscapes in the South; they also maintained a synthesis of modern and historical knowledge necessary for recreating such landscapes across the region and nation.

The idea for Tall Timbers began to take shape at a weekly gathering of friends. Stoddard had long been at the center of a remarkable community of naturalists surrounding Sherwood. Ed Komarek and his wife, Betty, purchased the neighboring property in 1938 and called it Birdsong. Ed's brother, Roy, a fine wildlife biologist in his own right, joined them on Birdsong around 1945 and became a guiding voice in the area for the next three decades. Just down the road at Tall Timbers Plantation was its owner, Henry Beadel, always more of a naturalist than a hunter. In addition, forester Leon Neel, a Thomasville native with a unique appreciation for all things wild, began working for Stoddard in 1950. He and his wife, Julie, moved into a former tenant house on Sherwood that year and quickly took an active role in this unusual, rural community. During the mid-1950s, the group routinely met for Sunday morning coffee at Birdsong, where an aging and heirless Beadel began to wonder what to do with his property.[5]

Beadel initially looked into placing the property in a trust, hoping to preserve its environmental characteristics in perpetuity. Stoddard was friends with Richard Pough, a veteran of the Audubon Society and the recently named president of the Nature Conservancy, whom they contacted about the legal mechanics of private land preservation. Pough gladly shared his experience, but they all hoped the land would become something more than a nature preserve. Three world-class biologists and a budding ecological forester lived within a five-mile radius of Tall Timbers, and its proximity to Florida

State University in Tallahassee could provide access to an institutional ally in establishing a research program. In addition, Stoddard and Komarek had become fast friends with Eugene Odum at the University of Georgia, who was conducting his own ornithological research at Ichauway Plantation, a hunting spread near Albany owned by Robert Woodruff, the president of Coca-Cola. These immediate contacts, not to mention the many government and academic ties that Stoddard and Komarek maintained throughout the nation, gave them a starting point for developing the research institution they had talked about for years.

On February 7, 1958, Tall Timbers Plantation officially became Tall Timbers Research Station. Beadel was its president, Stoddard its vice president, and Ed Komarek its secretary-treasurer; they recruited a handful of interested naturalists in the area to fill out a board of directors. Tall Timbers was a "station" only in name for its first several years; they did not even have a balance sheet until 1960, when Henry Beadle and his brother, Gerald, donated the seed for an endowment, and Komarek began actively recruiting annual contributions from other preserve owners. Most expenses, however, were out of pocket. According to Komarek, whatever research activity that anticipated "the creation of the foundation had been paid on a personal basis."[6] Indeed, Tall Timbers was a volunteer organization for the first several years of its life. Existing staff—Beadel, the Komareks, Stoddard, and Neel—conducted whatever Tall Timbers business they could as an extension of their preexisting pursuits. Such a scrappy beginning would become an important point of pride for those involved in Tall Timbers. As Komarek later recalled about the early days, "purely financial limitations need not necessarily be scientifically limiting. It has been said that sometimes it is an advantage to have limited funds and unlimited imagination."[7] Komarek certainly did not lack for imagination. From the beginning, Tall Timbers was a product of his vision, and he ran with the notion of creating a nongovernmental institution for ecological research, rooted in academic science, yet free of its institutional constraints and specialization.

Tall Timbers was personality driven from the beginning, and none was more important than Ed Komarek's. Though he had been an important influence among southeastern naturalists and resource professionals since his arrival in the Red Hills in 1933, Komarek always worked in Stoddard's considerable shadow. Tall Timbers gave him an outlet to play a more public role as scientist and conservationist, and he did not squander that opportunity. During the years following World War II, Komarek had managed to carve

Figure 13. J. L. "Cowboy" Stephens, Herbert Stoddard, and Ed Komarek (l–r) resting after a day afield, ca. 1950. Stephens was a plant breeder and agronomist at the University of Georgia's Coastal Plain Experiment Station in Tifton, and a frequent field companion of Stoddard and Komarek. After spending years as Stoddard's assistant, Komarek emerged to become one of the nation's most important advocates for modern fire ecology as the leader of Tall Timbers Research Station. Courtesy of Leon and Julie Neel.

out a living not unlike Stoddard's. His "Farm and Game" consulting business sustained him through the final years of the war, after the demise of the Cooperative Quail Study Association in 1943. The consulting work led to a job managing Greenwood Plantation in 1945, the Red Hills property owned by John Hay "Jock" Whitney, the celebrity scion of the New York Whitney family. Whitney was generous in his support of Komarek, giving him almost total control of Greenwood's eighteen thousand acres. Along with Roy Komarek, who comanaged the land base, Ed helped to sculpt Greenwood into one of the finest spreads of old-growth longleaf pine woodland left on the southern coastal plain. One of the goals of the Komarek brothers was to use Greenwood as a demonstration area for the ecological management of wildlife, agricultural, and forest resources. Ed had become a studious agriculturalist on his own land at Birdsong, and Greenwood gave him the opportunity to integrate productive farming within an ecological forestry and wildlife program. Stoddard managed the woodlands of Greenwood, marking and cutting timber with Neel throughout the 1950s, and they could eventually boast an impressive record of both economic and ecological productivity. Greenwood, then, came to represent Stoddard and Komarek's management philosophy in its most fully fledged form. By the time Tall Timbers was up and running, Greenwood was perhaps its most important asset—the land functioned as an outdoor laboratory for applied management, and equally important, served as a unique visual representation of the type of land management Tall Timbers sought to promote.[8]

The early years of Tall Timbers, however, started slowly. With most of the leadership still engaged in daily work, there was little opportunity to sketch out much of a research agenda or public role for the institution. Stoddard was the only staff member who was semiretired from his day job; he had left most of the forestry work in the hands of Leon Neel in 1956 and returned to his first passion, ornithology. An appealing research opportunity—one that helped shape the future possibilities for Beadel's property—cropped up around that time when a local broadcaster approached Beadel about building a television tower on Tall Timbers, one of the highest points between Tallahassee and Thomasville. Beadel initially resisted on aesthetic grounds, but Stoddard was curious about the effect such a tower would have on bird populations, as well as what it might reveal about bird migrations. Ornithologists had long known of significant bird mortality at tall structures, and a tower on Tall Timbers, Stoddard reasoned to Beadel, would provide an opportunity to conduct long-term research on the phenomenon in a place accessible and controllable. After hearing Stoddard out, Beadel reconsidered, and the tower went up in late 1955. Stoddard spent most every morning for the next ten years combing its grounds for dead birds, collecting over thirty thousand individuals and 170 different species. Though he repeatedly expressed the hope "that something can be learned that will lessen the slaughter at the thousands of similar obstructions," his study did not really explore ways to ward birds away from tall structures; its purpose was to supply data to those national organizations looking to curb kills.[9] It did, however, make Leon County "the best-documented migration locality in the state, if not the region," according to one chronicler.[10] In addition, the early days of the project stirred Henry Beadel's imagination, and ultimately convinced him that a research station was the right course to take for his property.[11]

The tower project became the fledgling station's first official research project, and, critically, helped attract the attention of the larger scientific community. Eugene Odum was a particularly important colleague interested in the tower project. Stoddard and Odum had known each other for at least ten years by the time the tower went up. They met in the early 1940s through the American Ornithologists' Union and the Georgia Ornithological Society and immediately began collaborating on a series of projects. A young zoologist at the University of Georgia when he met Stoddard, Odum was just beginning a career that would help develop the field of modern ecology. His revolutionary textbook, *Fundamentals of Ecology* (1953), the most forceful presentation yet of ecology as a field of its own, was still a few years away, but Stoddard already recognized a unique mind in Odum. Their first collaboration, along with several other ornithologists, was the *Preliminary Check-list and Bibliography of Georgia Birds*, an exhaustive guide to birds within the state's boundaries. The checklist eventually morphed into Thomas Burleigh's *Birds of Georgia*, a richly illustrated guide that Stoddard saw through the publication process and helped to fund. More importantly, the checklist project joined Odum and Stoddard, two of the finest observers of the natural world in the South.[12]

Odum and Stoddard conducted extensive correspondence on the minutiae of birdlife while compiling the checklist, and also became frequent field companions when Odum made off to Ichauway Plantation during the summers.[13] Odum was affiliated with the Emory Field Station at Ichauway, a unique research station established in 1939 to study malaria ecology. A joint endeavor of Emory University, the U.S. Public Health Service, and Ichauway owner Robert Woodruff, the field station had a wide-ranging mission to study the ecological factors of malaria endemic to the southern coastal plain. Melvin Goodwin, the broad-minded head of the station, pursued research on all aspects of malaria ecology—including bird malaria—and shortly after World War II he agreed to bring on Odum to study bird habits, habitat, and parasitology as they related to malaria. That agreement allowed Odum and his graduate students to expand their studies to bird ecology in general. As he did so often, Stoddard became a frequent guide to the area and reviewed a large percentage of the manuscripts that followed.[14]

Odum considered the tower study significant not only for its potential findings about tower mortality and migration habits but also for the abundance of dead birds it produced. Wildlife scientists, and ornithologists in particular, had long been wary of collecting living animals for research purposes, and as a result, they often had a hard time securing large numbers of dead animals during the postwar years.[15] In fact, this was one of the great benefits of Stoddard's effort. Stoddard carefully preserved every bird he picked up. He stored a large percentage of the birds in deep freeze for Odum and other researchers who needed them; he preserved thousands of skins to send off to museums, drawing on his earlier professional life as a taxidermist; and he also kept a number of skins to form the core of a Tall Timbers natural history collection. Odum's use of the birds resulted in important fat studies that determined the amount of energy migrating birds had stored and retained when they struck the tower, early evidence of his interest in energy flows. In addition, Odum secured a small grant from the National Science Foundation (NSF) in 1959 for use at the tower, marking Tall Timbers' first contact with the kind of large funding agency that postwar science increasing relied on.[16]

The relationship between Stoddard and Odum reveals much about changing methods in the biological sciences in the postwar era. Stoddard's place in the postwar scientific community was tenuous, and, by extension, so was that of Tall Timbers. Stoddard was a pioneer in wildlife science, and recognized as such, but he spent the latter half of his career developing innovative land management practices that, while empirical in nature, had less to do with rigorous science than achieving preconceived management goals. In fact, he had not produced much of what the scientific community would call research since *The Bobwhite Quail.* Neither had Ed Komarek, who had collected a tremendous amount of raw data on small mammals for decades, but rarely assembled it into a presentable form.[17] In the meantime, the biological sciences became "big science," dependent on large grants, collaborative networks of scientists, and increasingly sophisticated methodologies. The use of complex statistical and quantitative techniques was conventional practice in fields such as population biology and ecology by the time Tall Timbers came to be.[18] Stoddard and Komarek's training, on the other hand, was in a natural history tradition based on close observation, and they never had the professional incentives of academe to keep up with changing methods.

The top-down nature of big science, while capable at organizing large collaborative projects, was less adaptable to local social and economic customs than the natural history tradition. The NSF funding pleased Stoddard—preparing the tower grounds for bird collection was labor intensive and expensive, and Stoddard himself paid for the entire project until that point. But he also hoped to steer clear of any additional bureaucratic wrangling that the money might bring. He told Odum, for instance, that the "assistance that you can give our project here should be as free as possible of red tape to be attractive to us." In addition, the preparation and maintenance of the tower grounds required a substantial amount of labor, and NSF accounting caused Stoddard some concern about how to get the funds in the hands of African American workers. "Most of my negro help," Stoddard wrote, are "day laborers picked up morning by morning as we can get them. Though I can pay them once a week, they much prefer it at the end of each day worked, and checks are not very acceptable to them, so we will have to work out something whereby I can reimburse them according to local labor custom."[19] Such a clash between local custom and an extralocal funding source was little more than an inconvenience, but it does shed light on the execution of field science on the ground. Despite its rigid standards for research design, big science still depended on the organization of local labor, and the laborers themselves had some power to negotiate their terms based on many years of experience in the local labor market. It took intermediaries like Stoddard to decode local custom, something he was trained to do as a worker in the natural history tradition. Even though science was becoming more highly organized, then, it still proved malleable to local circumstances in its implementation.

Odum was well aware of where science was heading—he was actually leading the pack in the field of ecology—and he expressed some concern about Tall Timbers' background in natural history as it got underway. When Stoddard, for example, solicited comments on an essay manuscript that reflected a lifetime of observations about drought in the southern coastal plain, Odum offered a tough appraisal. The essay came on the heels of widespread wildfires during the drought of 1954–55, when fire from the Okefenokee Swamp spread to over 350,000 acres in southeast Georgia and northeast Florida.[20] Stoddard believed that the industrial management of pinelands that rimmed the swamp made those fires worse, and he set out to write a long essay explaining how they became so destructive. He focused most intently on the regulatory roles of fire and water in coastal plain ecology. Wherever water flowed, stood, or tarried the majority of the time, fire rarely visited; wherever water drained freely, fire held forth as the major determinant of species composition. Periods of long drought expanded fire's range; wet years restricted it. The arrival of twentieth-century fire suppression in the majority of fire's range—the uplands—had created a cauldron of fuels ready to explode during the periods of long drought. This was not a revolutionary thesis, but nor had many of its finer points been studied in detail. Stoddard's manuscript was not really an effort to do so, but it was a reflective effort to explain the evolutionary ecology of the region. And in the process, he hoped to convince those with power to influence forest management practices—industry and the state—that the regular use of fire would make the dry years easier on everybody.[21]

Odum applauded the effort but had some modernizing suggestions. His comments were two-fold: he wanted more hard data regarding energy inputs and outputs, and he wanted Stoddard to see the coastal plain as a system seeking to achieve balance through fire and water. This should come as no surprise to anyone familiar with Odum's work. With his 1953 textbook, Odum became the primary architect of ecosystems ecology, which built on the earlier climax ecology of Frederic Clements, who had argued that vegetative communities progress through predictable successional stages until they reach a final climax stage. Rather than Clements's unilinear model of succession, though, Odum conceived of an ecosystem as a container through which energy flowed, in and out, and sought to measure that energy in tangible terms. Perhaps his most interesting comments to Stoddard reflect his attempt to incorporate fire and other agents of disturbance into his ecosystems model. Years before the ecological community attacked his ecosystems model for lacking an acceptable explanation of natural disturbance, Odum wrote to Stoddard that because some systems "produce an excess of organic matter . . . such ecosystems change rapidly and are *stable only* if there are *periodic setbacks* in plant production, which allows consumers (man included) to catch up! This," Odum continued, "is the modern ecologists' *'balance sheet' approach* and says the same thing as you have said." Indeed, Stoddard had reproduced such "periodic setbacks" with fire for decades, but Odum wanted to see what this meant in hard, scientific terms. "Mostly," Odum wrote, "I'm thinking of how to get your astute observations to the attention of people who want 'figures' as well as arguments. . . . If you just add a little 'balance sheet' philosophy and a bit of quantitative data about a thousand more people will pay attention to your writings. You owe this to ecology and to conservation!"[22] In the simplest terms, Odum wanted Stoddard to quantify the amount of energy—in terms of rain, wind, fuel, heat, and so forth—that moved into and out of the system. Such measurements of energy, he argued, would demonstrate how ecosystems in the coastal plain sought balance, or equilibrium.

Odum knew very well that such quantification held little interest for Stoddard. His real intent was more likely to prod both Stoddard and Komarek toward modern science as they went about building a research staff. "In this connection," Odum wrote, "here is a *most important point* for future *Tall Timbers* research. . . . Describing natural changes is good, but will not *convince the hard-boiled engineer* who, *unfortunately*, is having increasing control in a crowded human society; so we must put nature in terms of *thermodynamics* so the engineer can be *educated*! You can't approach him with Natural History alone!"[23] A computational model of nature, one in which the numbers always balanced out, Odum believed, was the only way to rationalize a conservationist vision for the technocrats who controlled so much of the world. Odum was in the process of codifying his own thermodynamic view of nature, and he believed Tall Timbers could provide a useful forum to help spread the message. If only, he urged, they could recruit researchers trained in the techniques of measuring energy inputs and outputs, and steeped in a "balance sheet philosophy."[24]

Stoddard took Odum's critiques seriously, and understood the shortcomings of observation as a source of scientific explanation. But he and Komarek also sought to use their unique backgrounds to offer a broad critique of the biological sciences. Komarek was particularly concerned that professional specialization and its attending jargon threatened to isolate useful science from society at large. A strong dose of the natural history tradition, he argued, was actually just what conservation-oriented science needed. Remembering their experience with setting up the Wildlife Society, Stoddard and Komarek remained concerned about the applicability of science by the general public. On the state of conservation science, Komarek wrote, "there appears to be a large gap between the scientific studies now going on all over the world and the dissemination of this material to the 'public generally' in easily understood form. . . . Scientific journals have not only created closed communication circles, but with increasing specialization tighten the circle in ever-tighter loops."[25] The science was not much good, in other words, when only a handful of people understood what it meant. As a research station independent from academia, industry, or the state, Komarek considered Tall Timbers as in an ideal position to transcend the problem of specialization and present the public with knowledge it could use.

Despite being rooted in such an exclusive landscape, then, the mission of Tall Timbers was remarkably public-minded. Its charter was clear on this point: "This corporation is organized exclusively for public scientific and educational purposes and for no other purposes." To fulfill its intention, the charter mandated the station's leadership to carry out "educational work in such fields as wildlife management and the proper use of fire as a management tool . . . [and] to publish and distribute to the public generally any knowledge or information acquired as a result of such research, experiments and studies."[26] Unlike the Cooperative Quail Study Association, Tall Timbers had no members or subscribers, and it was not beholden to the preserve owners. All publications were freely distributed on request, because, "if the publications were sold, even at cost, they still would not reach the 'public generally.' Those of the public who would purchase them would be those individuals already interested, so that the effect would be a closed communication circle again."[27] They printed five thousand copies of each research and conference publication, and sent them to university and government libraries, state land management officials, interested landowners, and any other individual or institution that requested them. By 1975 Tall Timbers was sending materials to almost six hundred libraries in forty-seven countries on five continents.[28]

Before arriving at that point, though, Tall Timbers was about building institutional credibility for the first years of its life. The tower study was a good start, but was only of interest to a rather circumscribed group of ornithologists. The real intent of Tall Timbers was to promote ecological land management to the public, and they would need to stake a claim on something with wider value to achieve the kind of social and political significance they sought. In that regard, the new research station would make its largest public mark in fire ecology. This made sense, of course. The burning controversies of the past decades already made the southern coastal plain a geographical center of extant fire discourse, and no one knew more about fire as both a natural process and a cultural tool than Stoddard and Komarek. More than that, though, the science of fire ecology would be at the heart of their critique of modern conservation.

The use of fire, more than any other conservation measure, exposed what both Stoddard and Komarek considered one of the shortcomings of environmental preservation, especially in regard to wild lands. They had argued for decades that any effort to protect land for its ecological value should be active in its management, an approach not so different from the applied sciences of agriculture and forestry. Komarek wrote that a guiding principle in the founding of Tall Timbers was "to place great emphasis on *habitat management*, so as to have better 'nature management' in place of the outmoded 'nature conservation' which has become virtually ineffective and meaningless."[29] By "nature conservation," Komarek meant letting nature take its course, the hands-off form of wilderness preservation that in essence meant no land management at all. Their experience in the longleaf pine woodlands led them to believe that such an approach actually diminished biological diversity. Not only that, it denied a human influence that was fundamental to many natural processes. "The words 'nature management,'" wrote Komarek, "are used in the context that man is a vital part of nature, not just an onlooker." He also recognized the utility of experience-based knowledge about local environments, and "that an understanding of natural principles and processes does not necessarily require formalized scientific processes."[30] The problem in the longleaf pine region, as Komarek understood it, was a lost collective knowledge about these natural processes, largely because the environments in which this knowledge developed were lost. Tall Timbers, then, was as much a project of knowledge restoration as it was ecological restoration. "Whatever terminology is used," Komarek insisted, "the underlying idea is to work toward an understanding of nature and her basic principles. Nature, to be commanded, must be obeyed; and surely in this case, understanding is the prerequisite for obedience."[31]

To create a new understanding of nature in the longleaf pine, one based in older convention but modified for the vernacular of modern conservation, Tall Timbers instituted its first annual Fire Ecology Conference in March 1962. They invited officials from nearby state forestry services, the U.S. Forest Service, the National Park Service, the Fish and Wildlife Service, as well as academic biologists, to assess the history, ecology, and application of fire not only in the coastal plain, but the nation. This was the first major gathering of its kind in any region, and according to Komarek, the first time the words *fire* and *ecology* were joined together in a public forum. Komarek defined fire ecology for conference participants as "the study of fire as it affects the environment and the interrelationships of plants and animals therein."[32] By this time, most officials agreed that fire had an economic place in coastal plain land management, but now the discussion turned more explicitly toward ecology.

This was a big step. For years, Stoddard and Komarek had attempted to direct the conversation about land management away from shear economics and toward ecology. They successfully wedded the two management goals in their own work, and the fire conferences offered a public outlet for resource managers with broader influence to speak about the ecology of fire. The Forest Service had directed the national agenda on fire for years—an agenda that was not much concerned with ecology—and now Tall Timbers provided "a major forum for an alternative vision of fire," as fire historian Stephen Pyne has put it.[33] Not only that, they joined the Forest Service at their own game of public relations. The Fire Ecology Conferences were to be as accessible as possible to the public. As Roy Komarek wrote in the preface of the first conference proceedings, "the public at large, the conservation groups, and the leaders of our educational systems must be re-educated to the concept that fire has a useful place and may even be a necessity in the conservation of some of our natural resources."[34] The conference itself was open to the general public, but more importantly, Tall Timbers would publish the proceedings unexpurgated and disseminate them widely.

The proceedings of the first conference, in both structure and tone, reflect the novelty of the enterprise, as well as the distinctiveness of what Tall Timbers had to offer. The presentations were split up into three sections—history, ecology, and application—and presenters represented a wide spectrum of natural resource professions and organizations. Roy Komarek's call for reeducation, despite the sinister overtones of the word, proved an important theme throughout the proceedings. Most everyone involved recognized that a southern fire culture was fast fading, and that state conservation initiatives had helped do it in. It was time, they argued repeatedly, to resurrect what knowledge remained and make it part of the public consciousness. Henry Beadel, Roland Harper, and Stoddard, the most senior participants, kicked off the event by drawing on their long experience in the woods and recalling a time when the permeating presence of wintertime wood-smoke just seemed de rigueur in the southern countryside. Beadel recollected a ritualistic scene earlier in the century when the "head Negro" at Tall Timbers "made a narration" in late winter to give tenants the go-ahead to begin burning. It was a winter ritual not unlike hog killing or night hunting: an economic and social activity with an ecological significance. After experimenting with fire suppression on a block of land, and seeing the resulting rough, Beadel concluded that "the old-time settlers knew *very well*, in fact better than some of us do now, what they were about when they burned yearly."[35]

Many conference participants, on the other hand, were newcomers to taking a fire culture seriously. For decades, anyone who publicly called for the use of fire in a public forum suffered indifference at best, but usually outright hostility. The tone of this conference, however, was different, and the former fire suppressors ate crow with good humor. R. A. Bonninghausen, the chief of the Florida Forest Service and an old hand at fire suppression, "felt I was in the wrong place," after listening to Stoddard and Roland Harper speak of their devotion to fire.[36] Another participant, Forest Service researcher Thomas Lotti, couldn't "help but comment on the fact that woods burning two years back was something like a dirty word, and woods burners were treated accordingly." At this conference, however, "so-called woods burners are meeting in convention under the name of fire ecologists."[37] Lotti's association of resource professionals with the rural culture of woodsburning was tongue in cheek, to be sure, but he was not as far off as he might have imagined. Participants possessed a great deal of folk knowledge about woodsburning, whether they had spent their professional careers promoting it or preventing it. In the course of eliminating a cultural practice such as burning, after all, you often have to know a lot about the practice. By including those who were once adversaries, Tall Timbers began to chip away at the political boundaries that prevented an open exchange of knowledge about a highly contested subject.

The research presented at the conferences also coincided with several important shifts in both science and government land management policy. The second gathering, in mid-March 1963, was of particular note. Just as the Komareks put the finishing touches on the program earlier that month, the National Park Service issued a report titled "Wildlife Management in the National Parks," better known as the Leopold Report. Drafted by the Department of Interior's Advisory Board on Wildlife Management, which was led by Aldo Leopold's son, Starker, the report insisted that the biotic communities of National Parks—"vignettes of Primitive America," in the report's language—were in need of active management, a goal-based notion of land stewardship that the founders of Tall Timbers could enthusiastically support. This was, in fact, the type of management that Stoddard and Komarek promoted for decades, though they did not always aim to recreate "the ecologic scene as viewed by the first European visitors," as did the Leopold Report. They were most excited about what the report had to say about fire. The advisory board suggested that "of the various methods of manipulating vegetation, the controlled use of fire is the most 'natural' and much the cheapest and easiest to apply."[38] As one might imagine, Stoddard was particularly fond of the Leopold Report, and ten days after its appearance, he triumphantly read passages from it to the participants of the second Fire Ecology Conference.[39] The Park Service was not alone in its about face. The Forest Service also continued to make policy changes within the context of the Tall Timbers conferences, and announced a shift in focus toward fire management, rather than control, at the 1974 Fire Ecology Conference in Missoula, Montana.[40]

The fire conferences, then, were big tent affairs. As a research station, Tall Timbers influenced land management in the southern coastal plain most deeply, but through its conferences became a leader in a national and global effort to reassess fire. Indeed, the conferences were practically the only nurturing place for a budding fire ecology movement. Participants gathered at Florida State University in Tallahassee until 1967, when Komarek—by now one of the primary names associated with the growing study of fire ecology—took the conference on the road to Lake County, California, ponderosa pine country. From there, the conference returned to Tallahassee intermittently, and also convened in New Brunswick, Texas, Oregon, and Montana. Komarek also organized a special conference in 1971 devoted to fire in Africa, and consistently invited speakers to discuss fire ecology in other regions of the world, including western Europe, the Mediterranean basin, the Middle East, Australia, and southeast Asia. The success of the Fire Ecology Conferences was rooted in demand: there were few other places for those interested in fire to congregate and publish. Furthermore, the new scientific discipline of fire ecology rotated on Tall Timbers' organizational axis. The Fire Ecology Conferences were a nongovernmental haven for researchers and practitioners to present, share, and collaborate, as well as an institution to disseminate knowledge to the public. In sum, the use of the Red Hills hunting preserves as a scientific laboratory, coupled with Herbert Stoddard's early convictions about fire in the longleaf pine and his persistent dissent from orthodoxy, eventually led to a global reconsideration of fire and its role in shaping wildlands.

In addition to providing a public forum on fire, Tall Timbers also began to engage in research of its own. While Stoddard had made many observational remarks about fire in *The Bobwhite Quail*, and had used and advocated for its use in the decades since, he had not really produced much modern research on fire ecology. Tall Timbers Research Station, though, had a mission to produce new knowledge about coastal plain ecology. Stoddard responded in late 1959 by designing a controlled field experiment consisting of eighty-four half-acre plots, each devised to burn at certain intervals—every year, every two years, every three years, and so on up to every seventy-five years. He also designated a season of the burn for each plot. Stoddard, Leon Neel, and Neel's brother-in-law Jimmy Greene marked off the plots, and with Melvina Trussell, a botanist at Florida State, took an inventory of all plant species. They spread the plots across a wide spectrum of topography, vegetative communities, and soil types, and evenly burned the land area in preparation for the study.[41] Such a long-term experiment would provide quantitative evidence about how vegetative communities changed or persisted under different fire frequencies. Perhaps more importantly, the visual evidence of different fire regimes, each within sight of the other, became a stark demonstration tool for visitors to the station. As Leon Neel remembered, "we knew what was going to happen to these plots for the most part, based upon lots of experience. But these experiments proved the point for those with less experience, and they gave our intuitive knowledge experimental form."[42]

Fire ecology was central to Tall Timbers' mission, but the founders soon began to expand their interests. Stoddard, the Komareks, Beadel, and Neel, for instance, entered the public debate over the imported red fire ant when the U.S. Department of Agriculture initiated an eradication campaign in the late 1950s. The issue created a great deal of hostility between agricultural and wildlife interests by the end of decade, and Tall Timbers did its part, initially to moderate the debate on the local level and later to sponsor a series of conferences on ecological pest control. This was not the first time Stoddard found himself in the middle of a debate over fire ants. During the CQI, he spent a great deal of time identifying predatory ants, and he became even more interested in ant predation and control during the CQSA years. Fire ant predation on quail was one of the few management problems that baffled Stoddard. He concluded in *The Bobwhite Quail* that ants could kill anywhere from 4 to 12 percent of quail hatchlings, but misidentified the most destructive species as the native thief ant (*Solenopsis molesta*).[43] No one, in fact, really understood what they were dealing with at the time. Identification of ants is notoriously difficult, and fire ants in the South were no exception. Experts constantly confused native and imported ants, which created further confusion about the behavior and preferred ecological niche of problem ants. Without such knowledge, control was near impossible.

To find some answers, Stoddard and Komarek hosted Harold Peters and Bernard Travis, investigators with the U.S. Bureau of Entomology and Plant Quarantine, from 1935 to 1938 to study the life history of ants in the Red Hills. Peters and Travis successfully sorted out much ant taxonomy and identified *Solenopsis germinata*—the tropical fire ant—as the only problematic ant for quail in the Red Hills. *Solenopsis germinata* was not native to the South, but it had easily found its place by this time. Like so many introduced plants and animals, it flourished in the disturbed environments of human settlement. It did particularly well in the open fire-maintained pinelands of the Red Hills, and resisted any method of ecological control. Even the repeated application of sodium cyanide to individual mounds had little effect except to disperse the colonies, resulting in an actual increase of colonies in a target area.[44] After several years of worry, Stoddard and Komarek finally decided the threat was overinflated. As Komarek recalled, "We had realized that the introduced fire ant (*S. germinata*) had become so well adjusted to the ecosystem here that this ant was no longer a material menace to quail in this region."[45] At the close of the 1930s, the Bureau of Entomology pulled its investigators from the Red Hills, and Stoddard was "not recommending any general control attempts against [ants] as a protection to nesting quail."[46]

Stoddard and Komarek continued to follow developments in fire ant research over the next two decades, and reacted with a certain amount of irritation when the issue of fire ant control cropped up again in the late 1950s. They knew very well how frustrating ant control could be and did not relish another round with those critters. This time, another imported fire ant, *Solenopsis invicta*, was the issue. *Solenopsis invicta* traveled from South America to Mobile, Alabama, most likely in the 1930s, and slowly made its way across the South, following the pathways of modern land disturbance. Like *Solenopsis germinata*, it thrived in the South's disturbed environments, but it was a much more aggressive colonizer. Accounts varied, however, on whether it constituted a major threat to agricultural, ecological, and human health, or was a mere nuisance. Whatever its actual effects, in 1957 the USDA began the "largest, longest, most costly insect eradication program in American history," according to historian Joshua Blu Buhs.[47]

Stoddard and Komarek had applied their share of poisons to fire ant mounds over the years, but the scale of the USDA campaign was something else altogether. How the imported red fire ant became the object of such an assault is a complex story beyond the scope of this chapter; suffice to say that entomologists, wildlife biologists, farmers, journalists, hunters, and ordinary citizens generated a cacophonous collection of research, hearsay, and opinion about its threat to the agricultural economy. And after many stops and starts at fire ant control, the USDA finally touted the postwar commercial revolution in synthetic chemicals as the savior of the southern countryside. The chemicals heptachlor and dieldrin—more powerful relatives of DDT—seemed particularly effective against the fire ant, and by 1957 the USDA devised a plan to use former World War II bombers to spray the chemicals across 20 million acres of the rural South.[48]

Fire ants were not the only animals susceptible to heptachlor and dieldrin, of course, and the community of wildlife conservationists played an especially important role in opposing their indiscriminant use. Stoddard and Komarek's old friend Harold Peters, now with the Audubon Society, was one of the more visible opponents of the USDA, and he kept the crew at Tall Timbers in the loop about the eradication plans. Not that they needed insider information; reports of large wildlife kills due to spraying quickly made the rounds among those interested in such things.[49] By late 1958 Stoddard was "cramming like a schoolboy with a test ahead, to assimilate the rapidly growing literature [on fire ants], attend meetings and visit experimental areas, and talk, talk, and more talk."[50] Stoddard's former business partner, Richard Tift, assembled a meeting of local agricultural and wildlife interests in the Albany region that year, and Tall Timbers held a series of local meetings as well, but they amounted to little more than local USDA officials deflecting criticism.[51]

Stoddard's views on the subject solidified as he learned more about the effect of broadcasting insecticides. The USDA's "attempted control (or 'eradication') by the broadcasting of deadly hydrocarbon poisons," he wrote in 1958, "may prove even more destructive to wildlife than the ants themselves." By 1961 he felt "that even heavy infestations of the pesky creatures is preferable, everything considered, to a poisoned countryside."[52] After the meetings yielded few results, Stoddard and Komarek decided to use their bully pulpit as local authorities to at least alert the local public in the Red Hills to the threat of a chemical countryside. Komarek had recently begun hosting a noontime television show—a blend of agricultural report and ecological forum—and had gained some local celebrity in the process. Stoddard was a frequent guest, and at one point in 1959, they had Harold Peters appear on the show to discuss fire ant eradication and its effects on wildlife.[53] Despite such effort, the campaign's operation was far beyond Tall Timbers' influence. But the cumulative outcry from local opponents and wildlife conservation officials, coupled with a restriction on the use of heptachlor in 1959 and Rachel Carson's publication of *Silent Spring* in 1962—which excoriated fire ant eradication as an assault on nature—brought the campaign to a standstill for a while. The program hung around in another form through the 1970s, but the fire ant proved too ecologically entrenched to die. Like *Solenopsis geminata* and many other species before it, *Solenopsis invicta* became naturalized in the southern environment.

The fire ant controversy alerted Tall Timbers to some of the emerging challenges to conservation in the postwar world, especially the growing dependence of American agriculture on synthetic chemicals. They were well equipped to shepherd an agrarian woodsburning tradition into the world of modern resource management, but the momentum and promise of the chemical control of natural processes proved difficult to combat. They did, however, provide a forum to discuss alternatives. The pest problems associated with large-scale agriculture, along with a growing unease with the pesticide solution, brought about a renewed discussion among entomologists and ecologists about the possibilities of biological pest control. The use of beneficial species to control the populations of harmful insects was a very old technique in agricultural practice, one nearly made obsolete by the sheer power of chemical control. Tall Timbers took biological control one step further to consider how their ecological approach to habitat management might reduce the reliance on chemicals. Toward that end, they began a series of conferences in 1969 called "Ecological Animal Control by Habitat Management." Willard Whitcomb, an entomologist at the University of Florida and a recent collaborator at Tall Timbers, most likely supplied Komarek with the inspiration for the conferences. Whitcomb was heavily involved in the biological control of agricultural pests and was in the midst of establishing what have come to be known as the "Whitcomb Plots" on Tall Timbers, a study designed to establish a harrowing and plowing schedule that attracted predators of agricultural pests.

Ed Komarek left little doubt at the first conference that the controversies over chemicals inspired the gathering. In his introductory remarks, he recited a litany of "pest" problems they had addressed in the Red Hills through habitat management, and concluded that "these various animal control problems have made us aware of the many complexities and interrelationships between the numerous components of the ecosystem. It has also been impressed upon us by these and other experiences that there is no simple answer to the control of obnoxious pests such as strictly chemical control."[54] Land managers, Komarek insisted, had little choice but to make value judgments about particular plant and animal species, because "one soon learns," when managing land, "that plant succession is something more than just a botanical term and animal pests can become personal as well as academic problems."[55] So pest control itself was not going away. But the chemical control of pests represented a narrow vision of nature, one that focused solely on the species in question without regard to its ecological connections. The use of chemicals not found in nature, in other words, ignored a vast store of knowledge about how plants, animals, and humans related to one another.

The habitat management gatherings provided a forum for the discussion of alternatives to outright chemical dependence but did not result in a broad reconsideration of pest control. The promise of chemical control proved too appealing, and the industry too powerful, to simply turn back the clock.[56] These conferences did, however, prod Komarek to think more deeply about what Tall Timbers had to offer other scientists and resource professionals. Ecology-based land management seemed their most important contribution, but finding an ecological baseline in a natural world of constant change presented a difficult dilemma. "Man should not forget that he is concerned with living things and that our knowledge about living things is limited," Komarek told conference participants. Beginning with a quote from his early mentor Warder Clyde Allee, he continued,

> "Life . . . is never complete; always there is change." Because of the nature of the process we call "life," biologists never have an exact point of reference to which they may relate their studies. Constant change inhibits such a reference. Furthermore, no two living things have ever been, or will ever be, exactly alike. Change and diversity are basic and universal principles of the natural laws that govern living organisms. It has been the disregard of these principles and natural laws that has evoked the need for environmental management.[57]

Such a realization—there is no point of reference in nature—could paralyze a scientist or land manager looking to make decisions about conservation landscapes. Indeed, conservation biologists still debate these questions.

Komarek based his solution on what he and Stoddard had done for decades. Their approach to land management rested on the artful application of ecological knowledge, a concept that had become key for everyone involved in Tall Timbers. "The fundamental goal of Tall Timbers," Komarek insisted, "is the application of ecological knowledge to the environment by manipulating habitat. . . . We accept the philosophy that man can live in harmony with natural laws only if he follows the intelligent use of ecological principles—environmental management." The application of ecological knowledge required something more than science; it also required experience and imagination. Tall Timbers' great contribution to conservation, Komarek hoped, would be to combine science and art: "In the words environmental management, two very different concepts are apparent. Ecology is the study of the ecosystem, which is a science, while the application of such knowledge is an art. . . . All [of us at Tall Timbers] have been directly involved in practicing the art of management and studying the science of ecology."[58] Humans, then, could live with nature, but only through the most cultural of constructions—art and science.

The first decades of Tall Timbers Research Station were a culmination of the work Stoddard began as a museum collector, continued as a wildlife biologist, and carried on as an ecological forester, all the while staying close to his roots as an ornithologist. Ed Komarek, Roy Komarek, Leon Neel, and many others carried Stoddard's vision of land management forward into the 1970s and 1980s, continuing to conduct research and present conservation alternatives from their independent outpost in the Red Hills. Ed Komarek was a particularly dynamic leader. Under his guidance, Tall Timbers became one of the principal centers for the new discipline of fire ecology; it grew into an important site for research on the terrestrial environments of the southern coastal plain; and it was the home to a philosophy of land management replicated in few other places. The research station emerged from its roots in the natural history tradition to become a fully formed institution of modern science, with all of the advantages and limitations that come with such status.

CONCLUSION

Herbert Stoddard died on November 15, 1970. Though almost too poetic to be true, he is said to have passed with a copy of Aldo Leopold's *A Sand County Almanac* in his lap.[1] Apocryphal or not, it is appropriate to link their legacies together. Leopold did more than anyone during the interwar years to promote a new way of thinking about the American landscape, and Stoddard did more than anyone to translate that thinking into applicable land management. Together, they realized that the prospects for nature in the modern world came down to a series of human choices. With their approach to wildlife management, they both started with a rather simple proposal: decipher what habitat conditions benefit desirable animal species, and then maintain or recreate those conditions. This narrow recognition grew into the realization that in most American landscapes, ecological stability and biological diversity could only be restored and maintained through active management of the environment. Without consciously biocentric choices about nature, they argued, land would become biologically devalued.

The purpose of this book has been in part to tell the story of lost environmental knowledge within changing landscapes, and the attempts to restore both through the growth of conservation science. The conservation science that took shape in the Red Hills arose in the Progressive era from a convergence of social and environmental trends. Concerns about health, recreation, and social status; the growth of government conservation; and an increasing knowledge in the biological sciences converged in the southern longleaf pine woodlands to create a new way of using and thinking about land altogether. The role of fire; the intricate relationships between predator, prey, and habitat; the biological value of agricultural landscapes; and the careful harvesting of timber all formed pieces of an ecological puzzle, a puzzle that Herbert Stoddard spent a lifetime dismantling, decoding, and reconstructing. Stoddard came to the region in 1924 with a taxonomic background in identifying and classifying the biological parts of nature, and over the next several decades helped to fit those parts together into an ecological approach to land management.

As much as the shifting paradigmatic trends in the biological sciences, though, the particular natural processes of the longleaf pine woodlands shaped Stoddard's form of conservation science. It was a disturbance-prone, stochastic ecological system, and those traits were reflected in Stoddard's system of land management. His approach to conservation science was itself stochastic, always shifting in response to environmental behavior or the nuances of a particular place. Because of these nuances, he rarely embraced ecological theory or land management formulae. As much as he attempted to suspend landscapes in stasis, the only normative environment he ever found was one based in disturbance, uncertainty, and change.

Stoddard's location in the longleaf pine woodlands was also important because it was a distinctly southern landscape. This disturbance-prone environment developed alongside, and was clearly suited to, agrarian activity. The management practices of small- to medium-scale agriculture and open-range livestock grazing fit rather nicely into this natural system, and they were key ingredients in Stoddard's practice of conservation. In addition, the southern hierarchies of race and class buttressed the implementation of conservation science in the Red Hills. The management of land also required the management of people, and Stoddard's conservation science capitalized on the severe inequalities found in the Red Hills landscape. Conservation in the Red Hills, then, had a dark side, but it was not one that entirely restructured historical relationships with the land. Instead, it maneuvered from within the social, cultural, and economic conventions already in place.

On a broader level, Stoddard shaped his conservation science largely under the radar of state conservation, but only in collusion with more visible friends such as Aldo Leopold, Waldo McAtee, Paul Errington, and many others. It was no accident that an American biocentrism surfaced through wildlife management in the 1920s and 1930s. Hunting was wildly popular, and game species were in decline. The perspective of wildlife management, however, was slow to gain a foothold in state conservation. Instead, the Red Hills hunting preserves stepped in as the movement's first real field laboratory, and we can locate both the strengths and weaknesses of Stoddard's conservation science in this private landscape. Though he came of age during the waning years of Progressive-era conservation, and worked closely with government organizations throughout his career, his major contributions came through quasi- or nongovernmental activities. Indeed, I don't think his way of thinking about the land could have come from any other place. As Leopold recognized, Stoddard was free to experiment with nature outside of bureaucratic oversight, and he had remarkable success doing it. He actually grew leery of his occasional work for the state, once telling Eugene Odum that "I have wasted months in the aggregate preparing reports on [state lands], and most of them have been buried in the files without doing a speck of good that I could ever see. These agencies just go ahead and do as they d——m please every time due to pressure from outside sources."[2] The land management goals of private hunting preserves, on the other hand, were largely in tune with his own, and they gave Stoddard an environmental space in which to animate his biotic view of land.

At the same time, Stoddard's growing resistance to working through well-organized public agencies was also a limitation of his conservation science. He was highly successful in circumscribed landscapes, but these were among the most exclusive landscapes in the nation. Leopold's early concern that Stoddard's conservation would "relapse into a mere private enterprise" did not fully come to fruition—Stoddard's work did, indeed, filter through hunting preserve boundaries, and became a foundation for the practice of wildlife management on both public and private land across the nation.[3] But the specter of wealth, privilege, and elitism has always shadowed his preserve landscape.

The organizational structure of the hunting preserves themselves also presented limitations for this nongovernmental conservation. Historians have largely understood the conservation and environmental movements from the perspective of liberalism; that is, improving and preserving the environment for the public good. But there is also a strong conservative strain in the origins of environmental conservation. In this case, the original landowners of the late nineteenth century were from the robber baron class that benefited most from the nation's industrial revolution, and their progeny continued to be among the nation's conservative vanguard. As the modern American state took shape in the early twentieth century, these landowners were concerned equally about the expanding role of government regulation and the groundswell of resistance to corporate power that challenged the social and economic status quo. Their retreat to the Red Hills and the type of conservation they supported was an important result of their conservatism. Despite Stoddard's role as an ambassador to public-oriented science and conservation, the landscapes themselves—though among the most biologically diverse in the nation—were closed to the general public. In their search for healthy landscapes, northerners looked for places to sequester themselves from the forces of modern, industrial America, but their form of conservation was less a reaction to the growth of industrial and corporate America than a concomitant to it. Just as did industrial agriculture and forestry, they did as they pleased under the property rights structure of the post–open range New South.

Since Stoddard's death, the landscapes of southern hunting preserves, and the coastal plain at large, have changed in a variety of ways. Modern forestry and agricultural practices continue to dictate many choices about land management, but other real estate pressures also present a challenge to conservation efforts in the coastal plain today. As in other areas of the nation, shifting demographic trends of recent decades have created new markets for land in former rural areas, namely commercial and residential development. For those who value a return on investment more than outdoor recreation or ecological stability—a common sentiment, it turns out—rural land is sometimes worth much more in houses or commercial outlets than in forestry, agriculture, and wildlife. Substantial blocks of longleaf pine woodland have continued to diminish as a result—down to a little over 2 percent of the former range at last count.[4] Even in the Red Hills and other hunting preserve regions, landownership patterns and land management goals lack the consistency they once had. The modern American custom of partible inheritance, wherein multiple heirs split an estate equally, has exposed a major weakness of conservation on these private landscapes. Large ownership blocks become split between heirs, some of whom rarely set foot on the place before selling their portion to one of various real estate interests. With the ignition of a diesel-powered engine, centuries of natural and cultural history can be erased in a day.

Some preserve owners, on the other hand, might love the sport of quail hunting too much, according to some land managers and ecologists. They emphasize managing for the bobwhite quail to such an extent that diversity suffers. Stoddard always maintained that one quail per acre was a reasonable target that would not diminish the ecological properties of the fields and forests. But today, many property owners aim for up to three wild birds per acre, a goal only attainable by liquidating woodlands and installing a seemingly endless series of feed patches and hedgerows. This trend—producing the maximum number of one particular species—is little different than fencerow-to-fencerow agriculture or short rotation pulp pine forestry. And I can't imagine that Herbert Stoddard would be pleased to see many of his management methods used for such a purpose. Produce too much of one thing, whether quail, cotton, or pine, and the rest inevitably suffers.

There are people better equipped than I, however, to critique modern conservation, and they are working hard to resurrect a serious interest in the ecological fate of the southern coastal plain. Longleaf pine is at the heart of that interest. The market for pulpwood has literally moved south, into the developing world where social, economic, and environmental conditions are ripe for just such an extractive industry. Because of the wood's high quality, several federal and state subsidies, and some say, a healthy nostalgia among older landowners for the tree, longleaf pine is now a viable option for landowners in the coastal plain. Large pockets of longleaf woodland still exist on hunting preserves, military bases, and national forests across the South, and they now serve as demonstration areas and inspiration for some landowners.[5] In addition, an assortment of federal and state conservation departments, academic researchers, and independent organizations such as the Longleaf Alliance, the Joseph W. Jones Ecological Research Center, and Tall Timbers Research Station, continue to manage important longleaf woodlands and offer technical assistance to interested landowners.

And Herbert Stoddard's former student, Leon Neel, continues to be a force in the land management and scientific circles of the southern coastal plain. Public and private land managers, forest ecologists, and practicing foresters now look at the old southern hunting preserves as a template for a new kind of conservation biology in the longleaf pine woodlands. It is a form of conservation practiced with an agrarian sensibility, a scientific rationale, and an artful vision. It is also practiced with an ecological consciousness that understands history—the knowledge that old-growth longleaf pine woodlands exist today only because people continuously made the choice not to cut them down. The choices of people, in other words, have something to do with the fate of nature, and the story of the Red Hills, though riddled with moral ambiguities, is also instructive of how to make some good choices.

NOTES

Introduction

1. *Dictionary.com Unabridged*, Random House, http://dictionary.reference.com/browse/manipulation.

2. The standard work on the development of natural resource conservation as public policy remains Hays, *Conservation and the Gospel*. On building a biological knowledge base, see Allen, *Life Science*; Rainger, Benson, Maienschein, *American Development of Biology*; Rainger, Benson, Maienschein, *Expansion of American Biology*.

3. See Warren, *Hunter's Game*; Jacoby, *Crimes against Nature*; Spence, *Dispossessing the Wilderness*. Also see Scott, *Seeing like a State*.

4. Judd, *Common Lands, Common People*.

5. On the shift to biocentrism in American environmental thought, see Worster, *Nature's Economy*; Dunlap, *Saving America's Wildlife*; Roderick Nash, *Wilderness and the American Mind*; Frederick R. Davis, *The Man Who Saved Sea Turtles*; Barrow, *Nature's Ghosts*. Scholars of Aldo Leopold's life and work have done much to reveal this shift as well. See Meine, *Aldo Leopold*; Meine, *Correction Lines*; Newton, *Aldo Leopold's Odyssey*; Flader, *Thinking like a Mountain*.

6. On the environmental history of the New Deal era, see Worster, *Dust Bowl*; Sutter, *Driven Wild*; Sutter, "Terra Incognita"; Maher, *Nature's New Deal*; Maher, "'Crazy Quilt Farming'"; Phillips, *This Land, This Nation*; Beeman and Pritchard, *Green and Permanent Land*; and Henderson and Woolner, *F.D.R. and the Environment*.

7. Hybrid landscapes—places where nature and culture meet—have recently become the subject of great interest to environmental historians, and one manifestation of that interest has been a renewed focus on agricultural landscapes. For an overview, see White, "From Wilderness to Hybrid Landscapes"; some of the best book length examples include, Fiege, *Irrigated Eden*; Langston, *Where Land and Water Meet*; Linda Nash, *Inescapable Ecologies*.

8. On work and nature, see Richard White, "'Are You an Environmentalist or Do You Work for a Living?': Work and Nature," in Cronon, *Uncommon Ground: Rethinking the Human Place in Nature*, 171–85; and Peck, "The Nature of Labor." On the value of local knowledge for science and conservation, see Scott, *Seeing like a State*.

9. For an extended discussion of how "the domination of nature can lead to

the domination of some people over others," see Worster, *Rivers of Empire*, 19–60, quote on p. 50. For an example from the South, see Giesen, "'Truth about the Boll Weevil.'"

10. Daniel, *Breaking the Land*, 4. On tenantry and the transformation of southern agriculture in the twentieth century, also see Kirby, *Rural Worlds Lost*; Fite, *Cotton Fields No More*; Aiken, *Cotton Plantation South*.

11. Stewart, "If John Muir Had Been an Agrarian," 147. Stewart has done more than anyone to situate the South within the historiography of environmental history, and vice versa. Also see his "Re-greening the South," "Southern Environmental History," "'Let Us Begin with the Weather?'" and *What Nature Suffers to Groe*, 1–20. Studies on the environmental history of the South have multiplied in recent years. For commentary on where the field is going and where it has been, see Sutter, "No More the Backward Region"; Morris, "More Southern Environmental History"; Graham, "Again the Backward Region?" Book length studies on the South include Cowdrey, *This Land, This South*; Silver, *New Face on the Countryside*; Silver, *Mount Mitchell*; Kirby, *Poquosin*; Kirby, *Mockingbird Song*; Carney, *Black Rice*; Glave and Stoll, *To Love the Wind*; Mikko Saiku, *This Delta, This Land*; Lynn A. Nelson, *Pharsalia*; Jack Davis, *Everglades Providence*; Giesen, *South's Greatest Enemy?*; Hersey, *My Work Is That of Conservation*; Strong, *Making Catfish Bait of Government Boys*.

12. On the role of place in the making of science, see Kohler, *Landscapes and Labscapes*; Kohler, *All Creatures*; Livingstone, *Putting Science in Its Place*; Shapin, "Placing the View from Nowhere."

13. On dispossession in the name of conservation, see Warren, *Hunter's Game*; Jacoby, *Crimes against Nature*; Spence, *Dispossessing the Wilderness*; Margaret Lynn Brown, *Wild East*; Pierce, *Great Smokies*; Megan Kate Nelson, *Trembling Earth*; Phillips, *This Land, This Nation*; Powell, *Anguish of Displacement*; Reich, "Re-creating the Wilderness."

14. Kohler, *All Creatures*, 8.

15. Ibid., 18.

16. On the rise of ecology, see Worster, *Nature's Economy*; Golley, *History of the Ecosystem Concept*; Hagen, *Entangled Bank*.

17. The following paragraphs rely most heavily on Jose, Jokela, and Miller, *Longleaf Pine Ecosystem*, but there are many places to look for background on longleaf ecology. For examples, see Wahlenberg, *Longleaf Pine*; Earley, *Looking for Longleaf*; Neel, Sutter, and Way, *Art of Managing Longleaf*; Brockway et al., *Restoration of Longleaf Pine Ecosystems*; Farrar, *Proceedings of the Symposium*; Longleaf Alliance Web site, http://www.longleafalliance.org/.

18. Robert K. Peet, "Ecological Classification of Longleaf Pine Woodlands," in Jose, Jokela, and Miller, *Longleaf Pine Ecosystem*, 51–93; U.S. Department of Agriculture, Natural Resources Conservation Service, *Soil Taxonomy*.

19. E. V. Komarek, "Fire Ecology—Grassland and Man," in *Tall Timbers Fire*

Ecology Conference, Proceedings, 4 (1965): 174; on the historical role of fire, also see Cecil Frost, "History and Future of the Longleaf Pine Ecosystem," in Jose, Jokela, and Miller, *Longleaf Pine Ecosystem*, 9–42.

20. On Native American and colonial land use in longleaf forests, see Silver, *New Face on the Countryside*; Krech, *Ecological Indian*; Etheridge, *Creek Country*.

21. D. Bruce Means, "Vertebrate Faunal Diversity," in Jose, Jokela, and Miller, *Longleaf Pine Ecosystem*, 157–213, esp. 191.

22. On stages of growth, see Dale G. Brockway, Kenneth W. Outcalt, and William D. Boyer, "Longleaf Pine Regeneration Ecology and Methods," in Jose, Jokela, and Miller, *Longleaf Pine Ecosystem*, 95–133.

23. Peet, "Ecological Classification."

24. Means, "Vertebrate Faunal Diversity"; Means, "Longleaf Pine."

25. Peet, "Ecological Classification."

26. Jose, Jokela, and Miller, *The Longleaf Pine Ecosystem*, 4.

27. On the complexities of ecological restoration, see Hall, *Earth Repair*; Hall, *Restoration and History*.

28. William Warren Rogers and Clifton Paisley have written most extensively about the region. See Rogers, *Antebellum Thomas County*; Rogers, *Thomas County*; Rogers, *Transition to the Twentieth Century*; Paisley, *From Cotton to Quail*; Paisley, *Red Hills of Florida*.

29. Botanist Roland M. Harper was one of the first to identify the Red Hills as a distinct geological region. See Harper, "Geography and Vegetation."

30. Gatewood et al., *Comprehensive Study*.

31. On the antebellum era in the Red Hills, see Baptist, *Creating an Old South*; Rogers, *Antebellum Thomas County*.

32. On the struggles of African Americans in the region after the war, see O'Donovan, *Becoming Free*.

33. Unlikely is the key word here. I do not mean to dredge up the earlier debates of human ecologists and anthropologists who argued that some societies reached a functional equilibrium with their environments. Perhaps the best-known and most controversial of these studies is Rappaport, *Pigs for the Ancestors*; for an overview see Worster, "History as Natural History," in *Wealth of Nature*, 30–44.

34. Those who do include Silber, *Romance of Reunion*; Starnes, *Creating the Land*; Starnes, *Southern Journeys*; and Youngs, "Lifestyle Enclaves."

Chapter 1. From Public Playground to Private Preserve

1. Triplett, "Thomasville, Georgia: The Health Resort," 37–38, in file 267, Elizabeth Hopkins Collection, Thomasville Geneological, History, and Fine Arts Library, Thomasville, Ga. (hereafter TGH&FAL).

2. *Tallahassee Weekly Floridian*, April 17, 1883, quoted in Paisley, *From Cotton to Quail*, 32.

3. Valencius, *Health of the Country*, 12.

4. The germ theory of disease, in particular, turned the physician's gaze inward to the body. Linda Nash writes, "Germ theory held that disease could be traced to singular and discrete etiologic agents that penetrated the body rather than to the much vaguer and more nuanced concept of imbalance. However, nineteenth-century medicine was intellectually capacious, and most physicians had no difficulty mixing germ theories with long-standing environmentalist beliefs" (*Inescapable Ecologies*, 49–50).

5. Mitman's work is most effective in making connections between concerns about health, natural resource use, and conservation. See his works *Breathing Space*, "In Search of Health," and "Hay Fever Holiday." Other works that have influenced my thinking about health in rural landscapes include Valencius, *Health of the Country*; Linda Nash, "Finishing Nature" and *Inescapable Ecologies*; and the entire volume of Mitman, Murphy, and Sellers, eds., *Osiris* 19 (2004).

6. Roderick Nash, *Wilderness and the American Mind*.

7. On the antimodern impulse among northern visitors to the South, see Silber, *Romance of Reunion*, and Lears, *No Place of Grace*.

8. On acreage totals, see Brueckheimer, *Leon County Hunting Plantations*, 86. Throughout, I usually refer to these hunting retreats as preserves rather than plantations. Both terms are problematic, but I think the former slightly more accurate. These places hinged on the preservation of class and racial hierarchies and became ecological sanctuaries as well. The term "plantation," though it carries many different meanings that may apply to these hunting spreads, implies a highly organized system of labor control for the purpose of commodity production. Although the organization of labor on hunting preserves was based on the plantation model, it was not the dominant organizational feature and certainly not as rigidly enforced.

9. Ott, *Fevered Lives*, 32–33.

10. Feldberg, *Disease and Class*, 33.

11. On tuberculosis at the turn of the century, also see Sheila Rothman, *Living in the Shadow*; Ellison, *Healing Tuberculosis in the Woods*; Bates, *Bargaining for Life*.

12. On other southern resort destinations at the turn of the century, see Starnes, *Creating the Land*; Stewart, *What Nature Suffers to Groe*; Youngs, "Lifestyle Enclaves"; Dillon, "Reconstructing Aiken"; Aron, *Working at Play*; Cuthbert and Hoffius, *Northern Money, Southern Land*.

13. Brinton, *Guide-Book of Florida*, 115–16.

14. Ibid., 9.

15. Ibid. On health and miasmas in the antebellum South, see Numbers and

Savitt, *Science and Medicine*; Valencius, *Health of the Country*; Stewart, *What Nature Suffers to Groe*, 139–46.

16. Brinton, *Guide-Book of Florida*, 120.
17. Barbour, *Florida for Tourists*, 194–95.
18. Brinton, *Guide-Book of Florida*, 123.
19. Bill, *Winter in Florida*, 7.
20. On rail lines, see Rogers, *Thomas County*, 101–27.
21. Barbour, *Florida for Tourists*, 67–71.
22. Torrey, *Florida Sketch-Book*, 207.
23. Barbour, *Florida for Tourists*, 71.
24. Historical Census Browser, available at the University of Virginia, Geospatial and Statistical Data Center (2004): http://fisher.lib.virginia.edu/collections/stats/histcensus/index.html (retrieved April 2006); on Leon County demographic change, also see Brubaker, "Land Classification."
25. Baptist, *Creating an Old South*, 251; on the antebellum Red Hills, also see Rogers, *Antebellum Thomas County*; Paisley, *Red Hills of Florida*; and Paisley, *From Cotton to Quail*.
26. On the transition from slavery to freedom in southwest Georgia, see O'Donovan, *Becoming Free*.
27. *Thomasville Times*, April 5, 1873.
28. Quoted in Rogers, *Thomas County*, 135.
29. Quoted in ibid., 137.
30. On north–south collaboration, see Woodward, *Origins of the New South*, 107–41. That this effort to attract tourism was a town-based movement is significant. The influence of towns on southern life was a relatively new phenomenon in the rural New South. The transition of wealth to land and the resulting settlement of the countryside led to an exponential growth of towns throughout the South. Capital investment flowed toward population centers, and towns like Thomasville and Tallahassee were beneficiaries. See Ayers, *Promise of the New South*, 55–80.
31. Hopkins, "Pine Forests of Southern Georgia," 6–7.
32. *Atlanta Constitution*, April 21, 1882, quoted in Rogers, *Thomas County*, 441.
33. U.S. Bureau of the Census, *Tenth Census of Agriculture*, 3:109, 111.
34. Quoted in Kenneth Thompson, "Trees as a Theme," 517.
35. Quoted in ibid., 521. Morton's work in phrenology also formed a base for the scientific racism that proliferated later in the century.
36. Ibid. On the health-giving properties of eucalyptus, see Kenneth Thompson, "Australian Fever Tree in California"; Tyrell, *True Gardens of the Gods*, 69–70; and Linda Nash, *Inescapable Ecologies*, 72.
37. Cromwell, "Influence of Trees on Health," 165.

38. Ibid., 179.

39. On other physicians making this link, see Linda Nash, *Inescapable Ecologies*, esp. chap. 2.

40. Cromwell, "Influence of Trees on Health," 166.

41. On the turn-of-the-century timber industry, see chapter 3.

42. Cromwell, "Influence of Trees on Health," 173.

43. "F.W.R. in Chicago Hotel Reporter," in Triplett, "Thomasville, Georgia: The Health Resort" (Thomasville, 1891), p. 27, file 267, Elizabeth Hopkins Collection, TGH&FAL.

44. J. Wyman Jones to John Triplett, July 28, 1888, in Triplett, "Thomasville, Georgia: The Health Resort" (Thomasville, 1891), p. 26, file 267, Elizabeth Hopkins Collection, TGH&FAL.

45. "Thomasville, Georgia: The Great Winter Resort among the Pines" (Thomasville, 1898), file 268, Elizabeth Hopkins Collection, TGH&FAL.

46. "American Forestry Congress," *New York Times*, October 17, 1889.

47. Triplett, "Thomasville, Georgia: The Health Resort" (Thomasville, 1891), p. 21, file 267, Elizabeth Hopkins Collection, TGH&FAL.

48. On hotel construction, see Rogers, *Thomas County*, 131–54.

49. Torrey, *Florida Sketch-Book*, 207–8.

50. *Thomasville Daily Times-Enterprise*, January 25, 1891.

51. Triplett, "Thomasville, Georgia: The Health Resort" (Thomasville, 1891), p. 19, file 267, Elizabeth Hopkins Collection, TGH&FAL.

52. James A. Brandon to Harriet Jones Brandon, March 8, 1878, Brandon Collection, 77.33.142, Thomas County Historical Society, Thomasville, Ga. (hereafter TCHS).

53. Dr. D. S. and Harriet Jones Brandon to James A. Brandon, February 25, 1878, Brandon Collection, 77.33.141, TCHS.

54. Reprinted in Triplett, "Thomasville, Georgia: The Health Resort" (Thomasville, 1891), p. 25, file 267, Elizabeth Hopkins Collection, TGH&FAL.

55. "A Pleasant Winter Resort," *New York Times*, December 5, 1885, 5.

56. "In Sunny Thomasville," *New York Times*, March 6, 1892, 20.

57. E. L. Youmans, "Extract from letter to *Popular Science Monthly*," in Triplett, "Thomasville, Georgia: The Health Resort" (Thomasville, 1891), p. 22, file 267, Elizabeth Hopkins Collection, TGH&FAL.

58. Brinton, *Guide-Book of Florida*, 123; also see Aron, *Working at Play*.

59. Beard, *American Nervousness*, vi, 26; on neurasthenia, see Bederman, *Manliness and Civilization*; Lears, *No Place of Grace*; Gosling, *Before Freud*; Lutz, *American Nervousness, 1903*.

60. Brinton, *Guide-Book to Florida*, 118.

61. Bill, *Winter in Florida*, 175.

62. Ibid.

63. Brinton, *Guide-Book for Florida*, 118.

64. Rogers, *Thomas County*, 152–53.

65. Wyman Jones to John Triplett, July 28, 1888, in Triplett, "Thomasville, Georgia: The Health Resort" (Thomasville, 1891), p. 27, file 267, Elizabeth Hopkins Collection, TGH&FAL.

66. "Thomasville, Georgia: The Great Winter Resort among the Pines," (E. M. Mallette, 1898), p. 26, file 268, Elizabeth Hopkins Collection, TGH&FAL.

67. See Schuyler, *New Urban Landscape*; Rosenzweig and Elizabeth Blackmar, *Park and the People*; Jackson, *Crabgrass Frontier*, 73–102; Anne Whiston Spirn, "Constructing Nature: The Legacy of Frederick Law Olmsted," in Cronon, *Uncommon Ground*, 91–113.

68. On antebellum hunting in the South, see Proctor, *Bathed in Blood*. On the growth of sport hunting after the Civil War see Reiger, *American Sportsmen*; Marks, *Southern Hunting*; Giltner, *Hunting and Fishing*; and Tober, *Who Owns the Wildlife?* 41–60. On the cultural meanings of hunting in America, see Herman, *Hunting and the American Imagination*.

69. See Schmitt, *Back to Nature*; and Armitage, *Nature Study Movement*.

70. Isenberg, *Destruction of the Bison*, chap. 6; Bederman, *Manliness and Civilization*, 170–215.

71. Dunlap, "Sport Hunting and Conservation"; and Dunlap, *Saving America's Wildlife*; for a different view, see Reiger, *American Sportsmen*.

72. See Warren, *Hunter's Game*; Jacoby, *Crimes against Nature*; Isenberg, *Destruction of the Bison*; Hahn, "Hunting, Fishing, and Foraging.

73. Marks, *Southern Hunting*, 171; and Brueckheimer, *Leon County Hunting Plantations*, 86.

74. There is a strong southern literary tradition that makes this connection. See Marks, *Southern Hunting*, 42, 170–73; Babcock, *My Health*; on the social history of southern quail hunting, also see Prewitt, "Best of All Breathing"; Blu Buhs, *Fire Ant Wars*, 27.

75. Historical Census Browser, available at the University of Virginia, Geospatial and Statistical Data Center (2004): http://fisher.lib.virginia.edu/collections/stats/histcensus/index.html (retrieved April 2006).

76. U.S. Bureau of the Census, *Tenth Census of Agriculture*, 3:406.

77. Harper, "History of Soil Investigation," 37.

78. Hallock, *Camp Life in Florida*, 135–36.

79. Triplett, "Thomasville, Georgia: The Health Resort" (Thomasville, 1891), p. 24, file 267, Elizabeth Hopkins Collection, TGH&FAL.

80. Shepard Krech, "The Georgia Shooting Wagon and Other Rigs," in Humphrey and Krech, *Georgia-Florida Field Trial Club*, 102.

81. Herbert Stoddard Field Diaries, February 11, 1924, Herbert L. Stoddard Papers (hereafter HLS Papers), Tall Timbers Research Station Archives, Tallahassee, Florida (hereafter TTRS).

82. Undated Advertisement, file 10, Elizabeth Hopkins Collection, TGH&FAL.

83. For background on Hopkins see his "Historical Sketch of Sherwood Plantation, from the Red Man 1814 to 1934," in file "Sherwood Plantation, Abstract of Title;" and file "Our Vanishing Wildlife, with Notes on the Past and Present," in the Elizabeth Hopkins Collection, TGH&FAL.

84. On the Leon Hotel, see Henry Beadel Diaries, January 1, 1914, Henry Beadel Papers, TTRS. On Thomasville hotels leasing land see hotel pamphlets in Moore Collection, TCHS.

85. There is much literature on Flagler and Plant. Two good studies are Braden, *Architecture of Leisure*; and Standiford, *Last Train to Paradise*.

86. Quoted from *Savannah Press*, *Thomasville Weekly Times-Enterprise*, May 12, 1905.

87. "Thomasville, Georgia" (Thomasville Tourist Association, undated), box 3, Moore Collection, TCHS.

88. On the transformation of American tourism in general, see Sears, *Sacred Places*; and Aron, *Working at Play*.

89. H. W. Hopkins to S. E. Hutchinson, undated, box "Real Estate: Sale of," folder "Charles Thorne," H. W. Hopkins Papers, TCHS.

90. Range, *Century of Georgia Agriculture*, 169–74.

91. H. W. Hopkins to E. B. Eppes, March 15, 1916, box "Real Estate: Plantation," folder "Mason's Susina," H. W. Hopkins Papers, TCHS.

92. H. W. Hopkins to Charles H. Thorne, October 20, 1920, box "Real Estate: Sale of," folder "Charles Thorne," H. W. Hopkins Papers, TCHS.

93. *Thomasville Weekly Times-Enterprise and South Georgia Progress*, April 21, 1905.

94. L. S. Thompson to H. W. Hopkins, April 1, 1916, box "Real Estate: Plantation," folder "Sherwood Plantation," H. W. Hopkins Papers, TCHS.

95. Iamonia was renamed Forshala after Hutchinson sold out to Harry Payne Whitney. See Rogers, *Foshalee*.

96. L. S. Thompson to H. W. Hopkins, November 9, 1911, box "Real Estate, Sale of," folder "Shooting Rights," H. W. Hopkins Papers, TCHS.

97. H. W. Hopkins to S. E. Hutchinson, March 8, 1912, box "Real Estate, Sale of," folder "Shooting Rights," H. W. Hopkins Papers, TCHS.

98. S. E. Hutchinson to L. S. Thompson, February 12, 1914, box "Real Estate, Sale of," folder "Shooting Rights," H. W. Hopkins Papers, TCHS.

99. From deed copies in Hopkins Collection, box "Real Estate: Plantation," folder "H. E. Thompson," H. W. Hopkins Papers, TCHS; also see Paisley, *From Cotton to Quail*, 83.

100. There has been much detail written on the creation of these preserves and the personalities involved. See Paisley, *From Cotton to Quail*; Rogers, *Pebble Hill*; Rogers, *Foshalee*; Brueckheimer, *Leon County Hunting Plantations*.

101. H. W. Hopkins to E. B. Eppes, March 15, 1916, box "Real Estate: Plantation," folder "Mason's Susina," H. W. Hopkins Papers, TCHS.

102. See Wright, *Old South, New South*, 49.

103. Hahn, "Hunting, Fishing, and Foraging"; King, "Closing of the Southern Range"; McDonald and McWhiney, "South from Self-Sufficiency to Peonage: An Interpretation," *American Historical Review* 85 (December, 1980): 1095–118.

104. *Acts and Resolutions of the General Assembly of the State of Georgia*, 1876, 334–35. Steven Hahn ("Hunting, Fishing, and Foraging") makes a compelling argument that the curtailment of hunting during the spring and summer months had as much to do with getting labor in the fields as it did with the protection of game animals. On the quail plantations there is little evidence that the motives for enclosure extended beyond a concern for quail—a possessive concern, but a concern nonetheless.

105. Ned Crozer to H. W. Hopkins, June 15, 1912 (first quotation); Ned Crozer to H. W. Hopkins, July 26, 1912 (second quotation), box "Letters of Judge Hopkins," folder "Plantations, Not Typed," H. W. Hopkins Papers, TCHS.

106. On the range of quail, see Stoddard, *Bobwhite Quail*, 170; on ownership of wildlife, see Tober, *Who Owns the Wildlife?* 146–51; and Aldo Leopold, "Game Methods: The American Way," in *River of the Mother of God*, 156–63.

107. Bloch, *French Rural History*, 210. On European enclosure, also see E. P. Thompson, *Whigs and Hunters*; Blum, *End of the Old Order*.

108. See Hahn, "Hunting, Fishing, and Foraging," 37–39.

109. *Tallahassee Weekly True Democrat*, November 25, 1910.

110. Unknown sender to Hopkins, June 12, 1912, box "Read Estate: Plantation," folder "Mason's Susina," H. W. Hopkins Papers, TCHS; on hiring game wardens, see "Statement of Rents from Susina," folder "Mason's Susina," H. W. Hopkins Papers, TCHS.

111. Jas. C. Harvin to William P. Gainey, January 26, 1960, "Information on Mistletoe Plantation," file 140, Elizabeth Hopkins Collection, TGH&FAL.

112. On Grady County deer populations, see Jimmy Atkinson, interview with author, April 20, 2005, Stoddard-Neel Documentary Project, Forest History Society Archives, Oral History Collection, Durham, N.C. (hereafter FHS Archives).

113. John D. Archbold to H. W. Hopkins, June 24, 1924, box "Early Real Estate," folder "(Magnolia) Springwood #2," H. W. Hopkins Papers, TCHS.

114. Herbert Stoddard to Aldo Leopold, August 6, 1933, Aldo Leopold Papers, Series 9/25/10-1, box 3—Stoddard Correspondence, University of Wisconsin–Madison Archives, Steenbock Library, Madison, Wis. (hereafter AL Papers).

115. *Tallahassee Weekly True Democrat*, July 3, 1914.

116. *Tallahassee Semi-Weekly True Democrat*, January 21, 1913.

117. *Tallahassee Daily Democrat*, January 30, 1920.

118. On the legal distinction between tenant and sharecropper, see Range, *Century of Georgia Agriculture*, 85.

119. Beadel Diaries, January 26 and January 30, 1914, Henry Beadel Papers, TTRS.

120. Henry Vickers interview, 1971, by Bobby Crawford and Betty Ashler, TTRS. On tenant life on Tall Timbers, also see Bauer, "At Home"; and Hamburger, "On the Land for Life."

121. Rent List, box "Real Estate," folder "Mason/Archbold," H. W. Hopkins Papers, TCHS.

122. Paisley, *From Cotton to Quail*, 104–5.

123. See pamphlet "Among the Pines: An Ideal Winter Resort, Piney Woods Hotel," in Moore Collection, TGH&FAL.

124. "T. F. Whiteman from Latrobe, PA," *Thomasville Times Enterprise*, January 17, 1891; Field, *Bright Skies and Dark Shadows*, 99.

125. Stoddard, *Bobwhite Quail*, 366.

126. For more on the northern tendency to "view African Americans as simply another feature of the landscape," see Silber, *Romance of Reunion*, 78–82. There is a large scholarship on the picturesque and its tendency to obscure social and economic inequalities through "naturalization." For an introduction, see Raymond Williams, *Country and the City*; Bermingham, *Landscape and Ideology*.

127. For more on the adoption of the Old South myth by northerners, see Cobb, *Away Down South*, 67–78.

128. Quoted in Karson, "Warren H. Manning," 118.

129. "Mill Pond Plantation Report of Architects," in "Mill Pond Plantation, Some Early History," file 177, Elizabeth Hopkins Collection, TGH&FAL.

130. I detail the political economy of southern burning in chapter 3. Even Herbert Stoddard, whose field notes provide the most compelling evidence that fire was still common in the early 1920s, would write late in life that fire exclusion was the cause of the quail decline. See Stoddard, *Memoirs of a Naturalist*, 194; also see Komarek, *Quest for Ecological Understanding*, 18.

131. Raymond Williams, *Country and the City*; also see Bermingham, *Nature and Ideology*, 57–86; on nature and artifice in the English countryside, see Price, *Flight Maps*, 114–24.

132. On the concept of second nature, see Schmidt, *Concept of Nature in Marx*; Smith, *Uneven Development*; Cronon, *Nature's Metropolis*.

Chapter 2. The Development of an Expert

1. Much of this chapter is informed by historians of science, especially those who focus on the rise of modern biological science at the turn of the twentieth century. For starters, see Rainger, Benson, and Maienschein, *American Development of Biology* and *Expansion of American Biology*; Pauly, *Biologists and the Promise*; Kohler, *Landscapes and Labscapes* and *All Creatures*; Farber, *Finding Order in Nature*. On the development of American research universities, see Veysey, *Emergence of the American University*; Geiger, *To Advance Knowledge*; Kohler, "The Ph.D. Machine."

2. Kohler, *All Creatures*, 8.

3. Wiebe, *Search for Order*; Kohler, *All Creatures*, 17–30.

4. Barrow, *Passion for Birds*, 184.

5. On the growth of ecology, see Worster, *Nature's Economy*; Dunlap, *Saving America's Wildlife*; Golley, *History of the Ecosystem Concept*; Mitman, *State of Nature*.

6. Florida was one of the most substantial "inner frontiers" remaining in the East. See Kohler, *All Creatures*, 33.

7. Stoddard, *Memoirs of a Naturalist*, 36. Again, Robert Kohler's analysis of naturalist autobiography offers insight into Stoddard's remembrance of childhood. He calls autobiographical descriptions of the boy naturalist "a literary trope: a foretelling of the future scientist in the curious youth. But they also bear witness to experiences of actual places and of a particular time in the environmental history of North America. They are stylized recollections of what it felt like to inhabit the twilight zone, where wild (or wildish) nature was experienced through the ideals of town culture. So it is no surprise that the trope of the boy-naturalist especially appealed to the naturalists whose career happened to intersect with the inner frontiers. It was an element out of which scientific identities were constructed" (Kohler, *All Creatures*, 40).

8. Though it's uncertain which company acted as the Stoddards' agent, at least two Flagler-owned land companies were in the area: the Chuluota Land Company and the Model Land Company. See Robison and Andrews, *Flashbacks*.

9. Stoddard, *Memoirs of a Naturalist*, 5.

10. Stoddard, "Memories and Reflections," 13 (in author's possession). This is the unedited manuscript of *Memoirs of a Naturalist*, in which Stoddard was more candid and expansive with many of his views. Where the texts match, I quote from the published version; otherwise, I draw from the manuscript.

11. Ibid., 23.

12. On the commercial development of natural history, see discussion below, and Keith R. Benson, "From Museum Research to Laboratory Research: The Transformation of Natural History into Academic Biology," in Rainger, Benson, and Maienschein, *American Development of Biology*, 49–86; Barrow, "Specimen Dealer."

13. Tebeau, *History of Florida*, 273–92. On Florida cattle ranching, see Ackerman, *Florida Cowman*; Mealor and Prunty, "Open-Range Ranching"; Otto, "Open-Range Cattle-Herding" and "Open-Range Cattle-Ranching."

14. Stoddard, "Memories and Reflections," 52.

15. Mealor and Prunty, "Open-Range Ranching," 162–63.

16. Stoddard, "Memories and Reflections," 107.

17. Ibid., 53.

18. Ibid., 108.

19. Mealor and Prunty, "Open-Range Ranching," 363.

20. Quoted in Otto, "Open-Range Cattle-Herding," 322.

21. Stoddard, *Memoirs of a Naturalist*, 58

22. Stoddard, "Memories and Reflections," 116.

23. On the nature study movement, see Armitage, *Nature Study Movement*; Kohlstedt, *Teaching Children Science*; and Schmitt, *Back to Nature*.

24. Stoddard, *Memoirs of a Naturalist*, 64.

25. Ibid., 68.

26. Stoddard, "Memories and Reflections, 147.

27. On Sauk County, see Flader and Steinhacker, *Sand Country of Aldo Leopold*; Lange, *County Called Sauk*.

28. Janet Davis, *Circus Age*, 10.

29. Ibid., 148.

30. On finding the slug, see Stoddard, *Memoirs of a Naturalist*, 85. On American perceptions of Africa during the Progressive era, see Bederman, *Manliness and Civilization*; Haraway, *Primate Visions*.

31. Kohlstedt, "Nineteenth-Century Amateur Tradition."

32. Benson, "From Museum Research to Laboratory Research," 50.

33. See Veysey, *Emergence of the American University*; Geiger, *To Advance Knowledge*; and Kohler, "The Ph.D. Machine."

34. Barrow, "Specimen Dealer"; and Kohlstedt, "Henry A. Ward."

35. See Haraway, *Primate Visions*, 26–58; and Wonders, *Habitat Dioramas*, 134.

36. Stoddard, "Memories and Reflections," 170.

37. Stoddard, *Memoirs of a Naturalist*, 88. For Akeley's adventures, see Akeley, *Wilderness Lives Again*; and Bodry-Sanders, *Carl Akeley*.

38. Haraway, *Primate Visions*, 29. Haraway's trenchant cultural analysis also reveals the deeply ingrained race and gender hierarchies that Akeley's displays represented, and taxidermy's role in preserving a white male authority that was under threat from a decadent, effeminizing urban industrial order. My own analysis of taxidermy is less ambitious. As should become clear, I am more interested in taxidermy's ability to uncover hidden local nature, as well as the web of social interactions necessary to build a display, particularly in the field.

39. Stoddard, "Stuffed Birds"; Gromme, "In Memoriam."

40. Star, "Craft vs. Commodity."

41. Hornaday, "Common Faults," quoted in Star, "Craft vs. Commodity, Mess vs. Transcendence," 262.

42. Stoddard to S. C. Simms, September 22, 1918. box 12, folder 1, Conover Correspondence, Archives of the Field Museum of Natural History, Chicago, Ill.

43. For a discussion of the processes behind museum display see "Focus: Museums and the History of Science," 559–608; esp. Samuel J. M. M. Alberti, "Objects and the Museum," 559–71; and Sally Gregory Kohlstedt, "'Thoughts in Things': Modernity, History, and North American Museums," 586–601.

44. Stoddard, "Memories and Reflections," 209–11.

45. Ibid., 213.

46. Gromme, "In Memoriam," 873.

47. On creating representations of nature, see Sutter, "Representing the Resource."

48. On nature study and public education in Chicago, see Kohlstedt, "Nature, Not Books"; Rudolph, "Turning Science to Account."

49. Stoddard, "Memories and Reflections," 200.

50. Ibid., 201.

51. S. A. Barrett to R. S. Scheibel, April 12, 1922, Taxidermy Correspondence, 1921–1925, Manuscripts, Milwaukee Public Museum, Milwaukee, Wis. (hereafter MPM).

52. Ibid.

53. S. A. Barrett to George Shrosbree, July 13, 1922, Taxidermy Correspondence, Manuscripts, MPM.

54. Dorsey, *Dawn of Conservation Diplomacy*; see also Cioc, *Game of Conservation*.

55. S. A. Barrett to Herbert L. Stoddard, April 11, 1922, Taxidermy Correspondence, Manuscripts, MPM.

56. George Shrosbree, "Collecting Expedition to Bonaventure Island."

57. Stoddard Field Notes, p. 188, found in "Field Notes of Owen Gromme, April 1914 to July 1927," Manuscripts, MPM.

58. Mitman, *Reel Nature*; and Mitman, "When Nature Is the Zoo."

59. Stoddard, "Local Bird Notes."

60. Ibid. Clements's quote is from Clements, "Relict Method in Dynamic Ecology"; on "twilight zones," also see Kohler, *All Creatures*, 30–37.

61. Stoddard Field Notes, May 18, 1919, HLS Papers, TTRS.

62. Cowles, "Ecological Relations." Also see Kohler, *Landscapes and Labscapes*, 182–83; Engel, *Sacred Sands*.

63. Stoddard, "Memories and Reflections," 240.

64. Schantz, "Indiana's Unrivaled Sand-Dunes."

65. Ford to Herbert Stoddard, December 30, 1923, HLS Papers, Quail Investigation Correspondence, TTRS.

66. Gromme, "In Memoriam," 874.

67. Barrow, *Passion for Birds*, 172. Baldwin was married to Lillian Converse Hanna, sister of Mark and Howard Hanna, and was well-connected to the quail preserve set.

68. Stoddard, "Bird Banding in Milwaukee."

69. Between 1916 and 1923 Stoddard published ten times in *The Auk* and three in *The Wilson Bulletin*. For a full bibliography see Stoddard, *Memoirs of a Naturalist*, 285–88.

70. On the development of the Biological Survey, see Dunlap, *Saving America's Wildlife*; Cameron, *Bureau of Biological Survey*.

71. See Barrow, *Passion for Birds*, 172–75.

72. W. L. McAtee to Herbert L. Stoddard, January 2, 1924, W. L. McAtee Papers, box 10, Stoddard Correspondence, Manuscripts Division, Library of Congress, Washington, D.C. (hereafter LOC).

73. The backgrounds of the candidates for the job make clear that the Biological Survey had little idea of the methods and goals of the quail study. Stoddard's strongest competition came from Alden H. Hadley, a game propagator at a private game farm in Indiana who had little field experience, though he would go on to become an important voice in the National Audubon Society. See Alden H. Hadley to Herbert Stoddard, March 11, 1924, HLS Papers, Quail Investigation Correspondence, TTRS.

74. S. A. Barrett to Herbert Stoddard, March 23, 1922, Taxidermy Correspondence, 1921–25, Manuscripts, MPM.

75. E. W. Nelson to Herbert L. Stoddard, February 5, 1924, HLS Papers, Quail Investigation Correspondence, TTRS.

76. Milwaukee Public Museum Monthly Report, February 15, 1924, Manuscripts, MPM.

Chapter 3. Putting Fire in Its Place

1. Henry Beadel, "Fire Impressions," *Tall Timbers Fire Ecology Conference, Proceedings* 1 (1962):2–3.

2. Ibid., 3.

3. Stoddard Field Diaries, February 9, 1924, HLS Papers, TTRS.

4. Kohler, *Landscapes and Labscapes*, 6; also see Livingstone, *Putting Science in Its Place*; Shapin, "Placing the View from Nowhere."

5. Clifford Geertz argues that common sense is a cultural system and is possessed by "someone who is able to apprehend the sheer actualities of experience . . . [and] who is able to come to sensible conclusions on the basis of them." This is not to say Stoddard was exceptional; on the contrary, his common sense told him that the common sense of locals who lived and worked in the Red Hills would be an invaluable resource in the process of scientific discovery. See Clifford Geertz, "Common Sense as a Cultural System," in *Local Knowledge*, 76.

6. On the progression from game to wild life to wildlife management, see Etienne Benson, "From Wild Lives to Wildlife," unpublished essay (thanks to the author for providing me a copy of this essay); Meine, *Aldo Leopold*, 259; Newton, *Aldo Leopold's Odyssey*, 51.

7. Meine, *Aldo Leopold*, 264.

8. Aldo Leopold, "The Role of Universities in Game Conservation," *DuPont*

Magazine (June 1931), in AL Papers, 9/25/10-1, box 4, folder 3—Ralph King Correspondence.

9. Stoddard Field Diaries, February 12, 1924, HLS Papers, TTRS; also see H. W. Hopkins, "Historical Sketch of Sherwood Plantation, From the Red Man 1814 to 1934," file 177, Elizabeth Hopkins Collection, TGH&FAL.

10. Handley's relationship with Stoddard was an interesting one. Handley and his wife lived for a time in one of several small houses on Sherwood, but they never quite adjusted to country life and eventually moved to Thomasville. Handley stayed in the region until 1928 and completed important work on quail food habits, but he never adjusted to Stoddard's work habits. Stoddard, in turn, was little impressed with Handley's work and refused to write a recommendation at the conclusion of the study. Stoddard explained, bluntly, that Handley "failed to appreciate the opportunity presented and [was] inclined to regard your work as a 'Government job.'" Handley did secure a job in 1929, however, with W. B. Coleman at the Virginia State Game Farm and wrote up his food studies chapter from there. See Stoddard to Chas. O. Handley, September 14, 1928, HLS Papers, Early Correspondence A–Z, TTRS.

11. S. Prentiss Baldwin to Herbert Stoddard, March 11, 1924, HLS Papers, Baldwin Correspondence, TTRS.

12. S. Prentiss Baldwin to Herbert Stoddard, March 15, 1924, ibid.

13. On Thompson's background, see *New York Times*, March 26, 1936, 16.

14. Baldwin to Stoddard, March 15, 1924, HLS Papers, Baldwin Correspondence, TTRS.

15. Ibid.

16. Stoddard Field Diaries, February 10, 1924, HLS Papers, TTRS.

17. Ibid., September 26, 1924.

18. Stoddard, *Bobwhite Quail*, 195.

19. Stoddard Field Diaries, February 9, April 19, 1924, HLS Papers, TTRS.

20. On rattlesnakes and overshooting, see ibid., April 13, 1924; on house cats, September 26, 1924.

21. Ibid., January 4, 1925.

22. Stoddard to E. W. Nelson, June 10, 1926, HLS Papers, Investigation Correspondence, TTRS.

23. Stoddard Field Diaries, May 14, 1924, HLS Papers, TTRS.

24. On the earliest uses of carrying capacity, see Young, "Defining the Range"; and Dunlap, *Saving America's Wildlife*, 65–83.

25. Stoddard Field Diaries, April 3, 1924, HLS Papers, TTRS.

26. Stoddard to R. L. Ireland, May 20, 1957, HLS Papers, R. L. Ireland Correspondence, TTRS.

27. This chapter owes a great deal to the pathbreaking scholarship on fire by Stephen J. Pyne, as well as the earlier work of Ashley Schiff. When the Forest

Service was still debating the role of fire, Schiff had already detailed the early struggles over fire within its ranks, and gave much attention to the fire debates that originated in the longleaf forests. See Schiff's *Fire and Water*. Pyne's work is crucial to anyone wanting to understand the political, cultural, and physical aspects of fire. A good place to start is Pyne's *Fire in America* (quote on p. 145).

28. See Silver, *New Face on the Countryside*, 59–63; Pyne, *Fire in America*, 145–55; Prunty, "Some Geographic Views of the Role of Fire in Settlement Process in the South, *Tall Timbers Fires Ecology Conference, Proceedings* 4 (1965): 161–68; Earley, *Looking for Longleaf*, 73–131; Jack Temple Kirby, *Countercultural South*, 33–56.

29. Stoddard Field Diaries, February 11, 1924, HLS Papers, TTRS.

30. Ibid., February 11, May 18, 1924.

31. Ibid., March 7, 1925.

32. Ibid., March 11, 1925.

33. Ibid., March 1, 1925.

34. Ibid., May 6, 1924.

35. Herbert L. Stoddard, "Report on Cooperative Quail Investigation," 53.

36. Ibid., 53.

37. Ibid., 54.

38. Ibid., 56.

39. Greeley, "Relation of Geography to Timber Supply," 7. On land policy in the South, see Gates, "Federal Land Policy"; Gates, "Federal Land Policies." On this era of lumbering in the South, see Michael Williams, *Americans and Their Forests*, 238–88; Ayers, *Promise of the New South*, 123–31; Jones, *Tribe of Black Ulysses*; de Boer, *Nature, Business, and Community*; Wetherington, *New South*, 99–138.

40. Kirby, *Rural Worlds Lost*, 45–46; Wetherington, *New South*; Wetherington, *Plain Folks' Fight*.

41. Various schemes involving cutover lands proliferated in the first decades of the twentieth century. See, for example, *Lumber Trade Journal* 74, no. 9 (November 1, 1918).

42. Ashe, "Long-Leaf Pine," 1.

43. Ibid., 2.

44. Pyne, *Fire in America*, 101. The first major public stand off between fire advocates and anti-fire foresters occurred over the light-burning controversy in California. As trained foresters moved into California to manage new public forests and parks around the turn of the century, they encountered settlers and timber owners using light fires to keep fuels under control. The two sides engaged in a public debate that resulted in more systematic methods of fire control. See Pyne, *Fire in America*, 100–112.

45. Ashe, "Long-Leaf Pine," 16. Also see Ashe, "Forests, Forest Lands"; and Ashe, "Forest Fires."

46. Vance, *Human Geography of the South*, 136.

47. See Demmon, "Twenty Years of Forest Research"; Wakeley, "Adolescence of Forestry Research.".

48. Eldredge, "Administrative Problems," 344.

49. Pyne, *Fire in America*, 190–98, 145–46, 199–218.

50. See chapter 6 for a deeper discussion of longleaf regeneration.

51. Demmon, "Silvicultural Aspects," 325.

52. Eldredge, "Administrative Problems," 344.

53. Hardtner, "Tale of a Root," 357–58. Of the Forest Service's resistance to criticism about fire, Pyne notes its relative youth as an agency, as well as the youth of its personnel. He wrote that the Service "could initiate controversy but was not responsive to criticism; it was too homogeneous by training and temperament, and too self-conscious about its newly won political and intellectual stature." Pyne, *Fire in America*, 191.

54. "The Southern Forestry Education Project, Final Report" (Washington, D.C: The American Forestry Association, n.d.), 4, in American Forestry Association Papers, box 13, Reports—Southern Forestry Education Project, FHS Archives; also see Pyne, *Fire in America*, 169–71.

55. "Southern Forestry Education Project, Final Report," 5.

56. Ibid., 24.

57. W. B. Greeley to E. W. Nelson, June 10, 1926, HLS Papers, Fire Correspondence, TTRS.

58. E. W. Nelson to W. B. Greeley, June 17, 1926, ibid.

59. E. W. Nelson to Herbert Stoddard, June 18, 1926, ibid.

60. Stoddard, *Bobwhite Quail*, 402–3.

61. Ibid., 401.

62. Stoddard to R. M. Harper, January 25, 1932, HLS Papers, Fire Correspondence, TTRS.

63. Stoddard to S. W. Greene, May 8, 1931; Stoddard to Alfred O. Gross, March 11, 1931, ibid.

64. Stoddard to R. M. Harper, January 25, 1932, ibid.

65. Memo to Technical Committee No. V, National Land Use Planning Committee, June 24, 1932, AL Papers, 9/25/10-2, box 5, folder 11.

66. Ibid.

67. HLS to Fred Morrill, July 24, 1932, AL Papers, box 5, folder 11—National Land Use Planning Committee.

68. See chapter 5 for a fuller discussion of the public need for intentional game management.

69. HLS to Fred Morrill, July 24, 1932, AL Papers, box 5, folder 11—National Land Use Planning Committee.

70. Herbert Stoddard to H. H. Chapman, October 7, 1931, HLS Papers, Fire Correspondence, TTRS.

71. Roland Harper, "Geography and Vegetation," 182.

72. Ibid., 182–83; For more of Harper's writings on fire, see Harper, "Relation of Climax Vegetation"; Harper, *Economic Botany of Alabama*; Harper, "Forest Resources of Alabama"; Harper, "Natural Resources"; and Harper, "Fire and Forests." Harper also gave credit to Ellen Call Long of Tallahassee for her early recognition that the exclusion of fire would eventually convert a longleaf-grassland forest into an upland hardwood hammock. See Long, "Forest Fires." For more on Harper's life, see Shores, *On Harper's Trail*.

73. Roland M. Harper to Herbert Stoddard, January 21, 1932, HLS Papers, Roland Harper Correspondence, TTRS.

74. Stoddard Field Diaries, February 27, February 28, 1929, HLS Papers, TTRS.

75. "Report on Trip to Vinita, Oklahoma and Points in Mississippi, Georgia, and Florida during November 1929," Research Reports, 1912–51, box 20, part 2, RG 22, National Archives Records Administration, College Park, Maryland (hereafter NARA-II).

76. Hardtner, "Henry E. Hardtner," 35.

77. See also Chapman, "Some Further Relations of Fire."

78. Chapman, "Factors Determining Natural Reproduction," 8.

79. S. W. Greene to Ovid Butler, September 14, 1931, HLS Papers, Fire Correspondence, TTRS.

80. S. W. Greene, "Forests That Fire Made," 583.

81. Ibid., 584.

82. H. H Chapman to S. W. Greene, September 10, 1931, HLS Papers, Fire Correspondence, TTRS.

83. H. H. Chapman to Henry Hardtner, September 10, 1931, ibid.

84. Ibid.

85. HLS to H. H. Chapman, October 7, 1931, ibid.

86. HLS to Ovid Butler, September 26, 1931, ibid.

87. Hardtner, "Tale of a Root," 357; for the entire panel presentation see the same issue of the *Journal of Forestry*, 320–60.

88. Hardtner, "Tale of a Root," 360.

89. Herbert Stoddard to Aldo Leopold, February 18, 1935, AL Papers, Series 9/25/10-1, box 3—Stoddard Correspondence. Leopold did, indeed, have an interest in the natural role of fire. His views on fire were less than certain during his earlier years as a forest ranger in the Southwest—he had written of fire's role in preserving grasslands and controlling erosion, but had also called it "the enemy of the wild." Leopold's understanding of fire stabilized with time, and he was always a great supporter of Stoddard's findings. See Newton, *Aldo Leopold's Odyssey*, 67–70, 367n107; Meine, *Aldo Leopold*, 399.

90. HLS to W. L. McAtee, October 30, 1936, HLS Papers, Wildlife Society Correspondence, TTRS.

91. Herbert L. Stoddard, "Use of Fire on Southeastern Game Lands" (1935) in Stoddard, Beadel, and Komarek, *Cooperative Quail Study Association*, 47.

92. Stoddard to Roland Harper, June 27, 1943, HLS Papers, Roland Harper Correspondence, TTRS.

93. Stoddard, *Bobwhite Quail*, 411.

94. The Forest Service research at the Southern Forest Experiment Station resulted in the first major attempt to synthesize information on longleaf. See Wahlenberg, *Longleaf Pine*. See chapter 6 for a fuller discussion of this work.

95. Stoddard, "Use of Fire," 51.

96. Ibid.

97. Ibid., 49.

98. HLS to E. A. Schilling, November 21, 1935, HLS Papers, Game Possibility Reports, TTRS.

99. Stoddard's approach to management and discovery, though far less systematic and ambitious, shares a great deal with the relatively recent field of adaptive management. See Holling, *Adaptive Environmental Assessment and Management*; and Lee, *Compass and Gyroscope*.

Chapter 4. Stalking Wildlife Management

1. Dunlap, *Saving America's Wildlife*, 39; for more on the organization of predator campaigns within the Biological Survey, see Cameron, *Bureau of Biological Survey*, 46–65. Building on the work of Dunlap, the literature on predators has grown tremendously in recent years. For a sampling, see Mighetto, *Wild Animals*; Pritchard, *Preserving Yellowstone's Natural Conditions*; Young, *In the Absence of Predators*; Coleman, *Vicious*; Barrow, "Science, Sentiment, and the Specter of Extinction"; Manganiello, "From a Howling Wilderness."

2. Dunlap, *Saving America's Wildlife*, 48–70; on the Kaibab, also see Dunlap, "That Kaibab Myth"; Young, *In the Absence of Predators*.

3. On the history of ecology, see Worster, *Nature's Economy*; Hagen, *Entangled Bank*; Golley, *History of the Ecosystem Concept*; Kingsland, *Evolution of American Ecology*; on wildlife management, see Dunlap, *Saving America's Wildlife*; Steen, *Forest and Wildlife Science*; Meine, *Aldo Leopold*; Flader, *Thinking like a Mountain*; Newton, *Aldo Leopold's Odyssey*; Pritchard et. al., "Landscape of Paul Errington's Work."

4. On the environment as actor in the field sciences, see Kohler, *All Creatures*; Kohler, *Landscapes and Labscapes*; Blu Buhs, *Fire Ant Wars*; Schneider, "Local Knowledge."

5. On Errington's quail study, see Robert Kohler, "Paul Errington" (my thanks to the author for sharing an advance copy of this article); Pritchard, et al., "Landscape of Paul Errington's Work"; and Newton, *Aldo Leopold's Odyssey*, 134–36.

6. Treatments of the "pure" and "applied" sciences abound within the history

of science and technology. For a sampling regarding ecology, see Tobey, *Saving the Prairies*; Egerton, "History of Ecology, Part One" and "Part Two"; Young, "Defining the Range"; Young, *In the Absence of Predators*.

7. Stoddard Field Diary, February 7, 1924, HLS Papers, TTRS.

8. Typescript, "Our Vanishing Wildlife, With Notes on the Past and Present," by H. W. Hopkins, file 235, Elizabeth Hopkins Collection, TGH&FAL.

9. Stoddard Field Diary, February 7, 1924, HLS Papers, TTRS.

10. Ibid., February 10, 1924, TTRS.

11. Fisher, *Hawks and Owls*. Also see Barrow, "Science, Sentiment, and the Specter of Extinction," 73–74; and Evenden, "Laborers of Nature."

12. A. K. Fisher to HLS, July 26, 1926, HLS Papers, Investigation Correspondence, TTRS.

13. Ibid.

14. HLS to A. K. Fisher, August 7, 1926, ibid.

15. Meine, *Aldo Leopold*, 259–63.

16. Stoddard, *Memoirs of a Naturalist*, 218.

17. Stoddard, "Memories and Reflections," 453.

18. Stoddard, *Memoirs of a Naturalist*, 221; on Errington's background, see Kohler, "Paul Errington"; Pritchard et. al., "Landscape of Paul Errington's Work"; Schorger, "In Memoriam."

19. Herbert Stoddard, "Report on Trip to Michigan, Wisconsin, and Minnesota, January 10–February 4, 1930," Research Projects, 1912–51, box 20, part 2, RG 22, NARA-II.

20. Stoddard, *Bobwhite Quail*, 415.

21. Ibid., 416.

22. Ibid., 206.

23. Ibid., 207–8.

24. Ibid., 217.

25. Ibid., 211.

26. Ibid., 212.

27. Ibid., 224.

28. Ibid., 224–25.

29. Ibid., 226.

30. HLS to Paul Errington, August 27, 1935, HLS Papers, Errington Correspondence, TTRS.

31. Stoddard, *Bobwhite Quail*, 428.

32. Ibid., 420.

33. Ibid., 424.

34. Ibid.

35. "Harassed Hawks and Owls Found to Aid Farmers," *New York Times*, February 12, 1928, 136.

36. "Rod and Gun," *New York Times*, June 18, 1932, 18.

37. "The Last Word on the Bob-White Quail," *New York Times*, August 30, 1931, BR6.

38. Leopold, *Game Management*, 3.

39. Ibid., 230.

40. Ibid., 252.

41. HLS to Aldo Leopold, September 19, 1931, AL Papers, 9/25/10-6, box 6, folder 2.

42. Errington, "Vulnerability of Bob-White Populations," 111; McAtee quote is from McAtee, "Effectiveness in Nature."

43. Errington, "Over-Populations and Predation, 232." Also see Errington, "Northern Bobwhite."

44. Paul Errington to W. L. McAtee, May 18, 1933, W. L. McAtee Papers, box 21, Errington Correspondence, 1932–35, LOC.

45. HLS to Paul Errington, October 4, 1933, HLS Papers, Paul Errington Correspondence, TTRS.

46. HLS to Paul Errington, September 25, 1933, ibid.

47. Ibid.

48. Paul Errington to HLS, September 29, 1933, ibid.

49. Ibid.

50. HLS to Paul Errington, September 25, 1933, ibid.

51. Paul Errington to W. L. McAtee, December 14, 1937, McAtee Papers, Errington Correspondence, 1936–37, LOC.

52. The correspondence is not exact about their meeting place, only that they did indeed meet.

53. Paul Errington to HLS, November 23, 1937, HLS Papers, Paul Errington Correspondence, TTRS.

54. Errington and Stoddard, "Modifications in Predation Theory," 739.

55. Ibid., 737. For more commentary on this article, as well as Errington's more fully realized vision of predator–prey relations, see Errington, "Predator and Vertebrate Populations."

56. "Society of Wildlife Specialists," February 20, 1936, AL Papers, 9/25/10-2, box 9, folder 6.

57. Leonard William Wing, "Naturalize the Forest for Wildlife," 293.

58. W. L. McAtee to Aldo Leopold, August 5, 1936, HLS Papers, Wildlife Society Correspondence, TTRS.

59. W. L. McAtee to Paul Errington, November 30, 1936, McAtee Papers, box 21, Errington Correspondence, 1936–37, LOC [first quote]; W. L. McAtee to Aldo Leopold, August 5, 1936, HLS Papers, Wildlife Society Correspondence, TTRS [second quote].

60. The slightly altered name, Society for Conservation Biology, resurfaced

in 1985 as an international organization devoted "to advance the science and practice of conserving the Earth's biological diversity." Despite the anachronistic terminology, this goal is not so divergent from the original intent of the Wildlife Society. See Meine, Soulé, and Noss, "'A Mission-Driven Discipline.'"

61. For an introduction to the history of the Wildlife Society, see *Wildlife Society Bulletin* 15, no. 1 (1987): 1–152; and "The Wildlife Society: Its First Quarter Century," *Journal of Wildlife Management* 26, no. 3 (1962): 291–305.

62. Wallace Grange to Aldo Leopold, May 24, 1936, AL Papers, 9/25/10-2, box 9, folder 6.

63. See Aldo Leopold, "Wild Lifers vs. Game Farmers: A Plea for Democracy in Sport," and "Grand-Opera Game," in Flader and Callicott, *The River of the Mother of God*, 62–67, 169–72.

64. W. L. McAtee to Wallace Grange, June 3. 1936, AL Papers, 9/25/10-2, box 9, folder 6.

65. HLS to Aldo Leopold, W. L. McAtee, and Wallace Grange, May 30, 1936, AL Papers, 9/25/10-2, box 9, folder 6.

66. *Bulletin ESA* 4 (1923), quoted in Tjossem, "Preservation of Nature," 4.

67. Paul Errington to W. L. McAtee, November 23, 1936, McAtee Papers, box 21, Errington Correspondence, 1936–37, LOC.

68. Shelford helped to found the Ecological Union in 1946, which became the Nature Conservancy in 1950. See Croker, *Pioneer Ecologist*; Tjossem, "Preservation of Nature," 33–65.

69. HLS to W. L. McAtee, January 12, 1937, HLS Papers, Wildlife Society Correspondence, TTRS.

70. HLS to Walter P. Taylor, January 7, 1936, ibid.

71. HLS to W. L. McAtee, May 5, 1937, AL Papers, 9/25/10-2, box 9, folder 6.

72. HLS to Paul Errington, September 25, 1933, Paul L. Errington Papers, box 5, RS 13/25/51, Parks Library Special Collections, Iowa State University, Ames, Iowa.

73. E. V. Komarek to Walter Taylor, December 18, 1936, HLS Papers, Wildlife Society Correspondence, TTRS.

74. McAtee reported the vote count, 72–2, in W. L. McAtee to HLS, March 4, 1937, ibid. Ed Komarek recalled the snobbery accusation in E. V. Komarek to Gardiner Bump, February 6, 1939, AL Papers, 9/25/10-2, box 9, folder 9.

75. Errington recalled Leopold's displeasure with red tape in Paul L. Errington to W. L. McAtee, March 16, 1937, W. L. McAtee Papers, box 21, Errington Correspondence, 1936–37, LOC.

76. "Professional Training in Wildlife Work," November 30, 1938, pp. 1, 3, AL Papers, 9/25/10-2, box 9, folder 9.

77. AL to Rudolph Bennitt, September 8, 1938, ibid.

78. HLS to AL, August 23, 1938, ibid.

79. Ibid.

80. AL to Rudolph Bennitt, September 8, 1938, ibid.

81. For more discussion on Leopold's final statement, see Newton, *Aldo Leopold's Odyssey*, 275–78.

Chapter 5. Wild Land in Cultivated Landscapes

1. On Leopold and McAtee's plans for Stoddard, see AL to HLS, January 18, 1930; AL to HLS, February 11, 1930, AL Papers, 9/25/10-1—Stoddard Correspondence.

2. HLS to AL, February 5, 1930, ibid.

3. AL to HLS, February 11, 1930, ibid.

4. Fitzgerald, *Every Farm a Factory*.

5. Stoddard, *Bobwhite Quail*, 5.

6. Daniel, *Breaking the Land*; Kirby, *Rural Worlds Lost*; Fite, *Cotton Fields No More*; Aiken, *Cotton Plantation South*; Giesen, *Boll Weevil Blues*.

7. On the environmental history of the New Deal, see Worster, *Dust Bowl*; Sutter, *Driven Wild*; Sutter, "Terra Incognita"; Maher, *Nature's New Deal*; Maher, "'Crazy Quilt Farming'"; Phillips, *This Land, This Nation*; Beeman and Pritchard, *Green and Permanent Land*; and Henderson and Woolner, *F.D.R. and the Environment*.

8. Stoddard, Beadel, and Komarek, *Cooperative Quail Study Association*, 2–5, 67.

9. Herbert Stoddard to Paul Errington, August 27, 1935, Errington Papers, RS 13/25/51, box 5.

10. For an extended discussion of ecocultural landscapes, see Megan Kate Nelson, *Trembling Earth*, 1–10.

11. Stoddard, *Bobwhite Quail*, 342.

12. Stoddard commiserated with one correspondent about Talmadge's election, writing, "Yes, 'Old Gene' rode to victory on the 'nigger hate' wave in spite of all a lot of us could do" (HLS to Harold Sebring, July 23, 1946, Herbert L. Stoddard Papers, Plantation Correspondence, TTRS).

13. Stoddard, *Bobwhite Quail*, 6.

14. On the development of Leopold's views on agricultural lands, see Newton, *Aldo Leopold's Odyssey*; Sutter, *Driven Wild*, 89–98; Meine, "Farmer as Conservationist"; and Leopold, *For the Health of the Land*.

15. HLS to AL, April 3, 1930, AL Papers, 9/25/10-1, box 3—Stoddard Correspondence.

16. Aldo Leopold, "Report to the American Game Conference on an American Game Policy" [1930], in Flader and Callicott, *River of the Mother of God*, 150.

17. Meine, *Aldo Leopold*, 278. Information in this and the following paragraph comes from pp. 259–90, and from Sutter, *Driven Wild*, 91–92.

18. Aldo Leopold, "Report to the American Game Conference," 152–53.

19. Stoddard, "Game Production by Farmers," 1930, HLS Papers, Miscellaneous Manuscripts, TTRS.

20. W. L. McAtee to HLS, December 1, 1934, W. L. McAtee Papers, Stoddard Correspondence—1932–34, box 6, LOC.

21. Aldo Leopold to Joseph P. Knapp, September 18, 1930, AL Papers, R. T. King Correspondence, 9/25/10-1, box 4, folder 1.

22. Meine, *Aldo Leopold*, 282–84; Newton, *Aldo Leopold's Odyssey*, 117–23, 136–40; on Elton, also see Crowcroft, *Elton's Ecologists*.

23. AL to Charles Elton, November 12, 1931, private papers of Leon Neel.

24. HLS to AL, May 13, 1932, AL Papers, 9/25/10-1, box 3—Stoddard Correspondence.

25. Robert Goelet, a member of the CQSA, owned a French game preserve and sponsored Stoddard's trip.

26. Quote is from Herbert L. Stoddard, "General Report on Trip to France and England for Mr. R. W. Goelet, May and June 1935," HLS Papers, TTRS. For Elton's take on hedgerows, see Elton, *Ecology of Invasions*, chap. 7.

27. HLS to AL, June 14, 1935, AL Papers, 9/25/10-1, box 3—Stoddard Correspondence.

28. "Proposed Cooperative Quail Investigations and Demonstrations for the Bobwhite Quail Belt of the Southeast," HLS Papers, Southern Quail Demonstrations, TTRS.

29. Herbert L. Stoddard, "Report on LaGrange, GA, Type 2 Quail Demonstration Project," January 29–30, 1932, HLS Papers, Georgia Project, 1932, TTRS.

30. Wallace B. Grange and Ross O. Stevens, "General Report on the Experimental Quail Management Projects," October 12, 1932, HLS Papers, Southern Quail Demonstrations, TTRS.

31. Verne E. Davison to HLS, July 22, 1935, ibid.

32. In Flader and Callicott, *River of the Mother of God*, 181–202; Also see Meine, *Aldo Leopold*, 320–23; and Sutter, *Driven Wild*, 93–95.

33. HLS to Aldo Leopold, July 26, 1934, AL Papers, 9/25/10-1, box 3, Stoddard Correspondence; for Stoddard's comments on "The Conservation Ethic," see HLS to Aldo Leopold, November 8, 1932.

34. AL to HLS, November 5, 1932, AL Papers, 9/25/10-1, box 3, Stoddard Correspondence.

35. HLS to AL, November 8, 1932, ibid.

36. Leopold eventually understood Stoddard's strengths and weaknesses, largely because Stoddard told him what he would and would not participate in professionally. When Jay "Ding" Darling approached Leopold about taking over

as chief of the Biological Survey in 1934, he declined, but briefly wondered what Stoddard might bring to the job. He wrote Darling, "It boils down, in my mind, to a choice between policy-making and research. Washington is obviously the place to write policy, and there is certainly a chance in Washington to (as you say) 'get 48 states in motion, instead of one.' But getting them in motion and *keeping* them in motion are two different things. I have not yet found the man who could organize and also direct such work. Stoddard was perfect in the latter respect, but not the former. In short, there is at least a doubt in my mind whether bringing research to actual fruition in one state is not, at this moment, just as important as starting it in many states" (AL to Jay N. Darling, May 29, 1934, AL Papers, 9/25/10-8, box 1, folder 2).

37. On the social problems of conservation, see Warren, *Hunter's Game*; Jacoby, *Crimes against Nature*; Spence, *Dispossessing the Wilderness*; Hahn, "Hunting, Fishing, and Foraging,"

38. Historical Census Browser, http://fisher.lib.virginia.edu.collections/stats/histcensus/php.county.php.

39. U.S. *Census of Agriculture*, 1930, vol. 2, part 2, 522.

40. Ibid., 574–75, 586–87, 598–99.

41. Ibid., 522.

42. K. S. McMullen, "Narrative Report of the County Agent, March 15, 1937–November 30, 1937," quoted in Paisley, *From Cotton to Quail*, 103. McMullen stated that the fixed rate was one five-hundred-pound bale for every forty acres, but that number seems too low, especially considering his statement that tenants had a hard time escaping debt. As one reader of this manuscript stated, "if a farmer can't grow a bale of cotton on forty acres, then they might need to find another line of work."

43. Farm Program, September 1939, Robert Winship Woodruff Papers, box 134, folder—Ichauway Plantations, Inc., 1929–39, Manuscripts, Archive, and Rare Book Library, Robert W. Woodruff Library, Emory University, Atlanta, Ga. (hereafter MARBL).

44. Stoddard, *Bobwhite Quail*, 367.

45. Ibid., 421.

46. Stoddard Field Diaries, January 2, 1928, HLS Papers, TTRS.

47. See Trimble, *Man-Induced Soil Erosion*; Trimble, "Perspectives on the History of Soil Erosion Control"; Hart, "Loss and Abandonment of Cleared Farm Land." On southern soil erosion during the early republic and antebellum periods, see Stoll, *Larding the Lean Earth*; for a different view of early agriculture, see Carville Earle, "The Myth of the Southern Soil Miner: Macrohistory, Agricultural Innovation, and Environmental Change," in Worster, *Ends of the Earth*, 175–210.

48. Stoddard, *Bobwhite Quail*, 350–51.

49. Ibid., 351.

50. Ibid., 364.

51. Arthur B. Lapsley to HLS, May 3, 1931, HLS Papers, Arthur B. Lapsley Correspondence, TTRS.

52. HLS to Arthur B. Lapsley, May 17, 1931, ibid.

53. On velvet beans and corn, see Range, *Century of Georgia Agriculture*, 186; and Kirby, *Mockingbird Song*, 103, 338n17.

54. "List of Rules and Regulations, December 7, 1934," Robert W. Woodruff Papers, box 153, folder—Ichauway Plantations, Inc., Tenant Contracts, 1934–35, MARBL.

55. Dan Lilly Interview, July 22, 1992, 3–4, Ichauway Documentary Project, interview by Wiley Prewitt, Joseph W. Jones Ecological Research Center at Ichauway Archives, Newton, Ga.

56. Heinsohn, *Southern Plantation*, 59.

57. Brown and Hadley, *African-American Life*.

58. Ibid., 34.

59. Ibid., 34, 39.

60. Ibid, 34.

61. Both of these women probably deserve more treatment than I can provide here. Still spoken of in reverent tones, they represent the quintessence of paternal, or maternal, benevolence. Kate was the daughter of Howard Melville Hanna, and took over possession of Pebble Hill in 1901. "Miss Pansy" was born in 1897 and inherited the place in 1936. See Rogers, *Pebble Hill*.

62. Brown and Hadley, *African-American Life*, 33.

63. Irene Hudson Interview, August 4, 1992, p. 18, Ichauway Documentary Project, interview by Sally Graham and Wiley Prewitt, Joseph W. Jones Ecological Research Center at Ichauway Archives.

64. Farm Program, September 1939, Robert Winship Woodruff Papers, box 134, folder—Ichauway Plantations, Inc., 1929–39, MARBL.

65. "Leon Count Agricultural Agent Annual Narrative Report, 1938," quoted in Brubaker, "Land Classification," 114–15.

66. Dan Lilly Interview, 28.

67. Stoddard Field Diaries, June 29, 1925, HLS Papers, TTRS.

68. "General Impressions of the 1927 Quail Nesting Season," ibid.

69. Herbert L. Stoddard, "The Bobwhite Quail: Its Propagation, Preservation, and Increase on Georgia Farms" (Atlanta: State Department of Game and Fish, 1933), 365–67, in Stoddard, Beadel, and Komarek, *Cooperative Quail Study Association*, 343–417. Stoddard also wrote a similar pamphlet for the North Carolina and Florida departments of game and fish.

70. Stoddard, "Bobwhite Quail . . . On Georgia Farms," 357.

71. Stoddard, *Bobwhite Quail*, 362.

72. Ibid., 363.
73. Ibid.
74. Ibid., 374.
75. Herbert L. Stoddard, "Notes on a Visit to Melrose and Pebble Hill Plantations," September 21, 1933, HLS Papers, Georgia Plantations—Melrose, TTRS.
76. Phillips, *This Land, This Nation*, 107–32; Wooten, *Land Utilization Program, 1934 to 1964*.
77. Stoddard, *Bobwhite Quail*, 405.
78. Herbert L. Stoddard, "Report on Game Possibilities of Certain National Forest Units in Mississippi and Louisiana," November 3–11, 1935, pp. 5–6, Research Reports, 1912–51, box 20, part 1, RG 22, NARA-II.
79. Herbert L. Stoddard, "Report on Game Possibilities of Certain National Forest Units of South Carolina, North Carolina and Georgia, January 9–17, 1936," pp. 1–2, ibid.
80. Herbert L. Stoddard, "Report on Game Possibilities of Certain National Forest Units in Mississippi and Louisiana," November 3–11, 1935, p. 9, ibid.
81. Ibid., 6.
82. "The National Plan For Wild Life Restoration, pt. 2: Tentative Projects for Migratory Waterfowl and Upland Game," Office Files of J. M. "Ding" Darling, 1930–35, box 1, RG 22, NARA-II. This report came from FDR's Committee on Wildlife, known to many as the "Beck Committee." Consisting of Thomas Beck, Jay "Ding" Darling, and Aldo Leopold, the committee set out to develop a national land-use plan for wildlife, and in the process had severe disagreements over several proposals. See Meine, *Aldo Leopold*, 315–19.
83. HLS to W. L. McAtee, April 10, 1934, W. L. McAtee Papers, Stoddard Correspondence, 1932–44, box 36, LOC.
84. W. L. McAtee, "The Biological Survey in Relation to Game Management," HLS Papers, W. L. McAtee Correspondence, TTRS.
85. About staying with preserve owners on his many consulting trips, Stoddard once wrote to Wallace Grange that "I stop in little hotels wherever it is convenient (which I prefer to having some flunkey unpack your bag and lay out your night gown, and all the agony that goes with staying over as guests of some of these men)" (HLS to Wallace B. Grange, February 15, 1931, HLS Papers, Early General Correspondence, TTRS).
86. HLS to W. L. McAtee, November 27, 1934, W. L. McAtee Papers, Stoddard Correspondence, 1932–44, box 36, LOC.

Chapter 6. From Wildlife Management to Ecological Forestry

1. HLS to Henry Beadel, October 13, 1941, HLS Papers, Plantation Correspondence, TTRS.

2. Stoddard, *Bobwhite Quail*, 370.

3. "Sixth Annual Report of the Cooperative Quail Association," in Stoddard, Beadel, and Komarek, *Cooperative Quail Study Association*, 127.

4. See Hirt, *Conspiracy of Optimism*, 22–25; on the shifting meanings of "ecosystem management" in the Forest Service, see Langston, *Forest Dreams, Forest Nightmares*, 269–80.

5. Jerry Franklin, interview by author, January 23, 2006. Conservation scientists at the Joseph W. Jones Ecological Research Center at Ichauway have put a great deal of work into translating Stoddard's forestry method into a working model of ecological forestry. See McIntyre et al., *Multiple Value Management*; Mitchell, Hiers, O'Brien, Jack, and Engstrom, "Silviculture That Sustains"; Mitchell, Hiers, O'Brien, and Starr, "Ecological Forestry in the Southeast"; *Fire Forest*.

6. HLS to Aldo Leopold, October 5, 1936, private papers of Leon Neel.

7. HLS to Aldo Leopold, May, 1937, ibid. On Stoddard's move into real estate, also see Stoddard, Beadel, and Komarek, *Cooperative Quail Study Association*, 118–22.

8. Hirt, *Conspiracy of Optimism*, 44–46.

9. HLS to Jack Jenkins, March 7, 1942, HLS Papers, Plantation Correspondence, TTRS.

10. HLS to Allen M. Pearson, October 17, 1942, HLS Papers, Alabama Cooperative Unit, TTRS.

11. E. V. Komarek to Hendon Chubb, May 19, 1943, E. V. Komarek Papers, Hendon Chubb Correspondence, TTRS; HLS to Owen Gromme, March 29, 1943, HLS Papers, Gromme Correspondence, TTRS.

12. Wakeley, "Adolescence of Forestry Research," 140.

13. Chapman, "Management of Loblolly Pine," 10.

14. Wakeley, "Adolescence of Forestry Research," 144.

15. Wahlenberg, *Loblolly Pine*; Wahlenberg, *Guide to Loblolly*; Schultz, *Loblolly Pine*.

16. The Forest Service released subregional results of the survey in phases throughout the 1930s. See William Clarence Boyd, "New South, New Nature," 52–54; and Maunder, *Voices from the South*.

17. Much of the following debate took shape after the publication of H. H. Chapman's "Management of Loblolly Pine." For reactions to this bulletin and Chapman's rebuttal, see Bull and Reynolds, "Management of Loblolly Pine"; Chapman, "Common Sense Needed"; and G. A. Pearson, "Intensive Forestry Needed," all found in the same issue of *Journal of Forestry* of October 1943. On even-aged and uneven-aged management in other regions, particularly the West, see Langston, *Forest Dreams, Forest Nightmares*; and Hirt, *Conspiracy of Optimism*, 136–39.

18. Chapman, "Management of Loblolly Pine," 17, 62.

19. Ibid., 61–62.
20. Ibid., 17.
21. HLS to Aldo Leopold, July 16, 1942, private Papers of Leon Neel.
22. Wahlenberg, *Longleaf Pine*, xi.
23. Wahlenberg, *Longleaf Pine*, 136.
24. Ibid., 100.
25. Ibid., 115–16.
26. Forest ecologists have since refined Wahlenberg's discussion of longleaf regeneration, of course, especially in regard to capturing a seed fall and securing reproduction. But an extensive knowledge is very much in evidence in Wahlenberg's work. For a sampling of more recent treatments, see Jose, Jokela, and Miller, *Longleaf Pine Ecosystem*; Brockway et. al., *Restoration of Longleaf Pine Ecosystems*; Brockway et. al., *Uneven-Aged Management*; Farrar, *Proceedings of the Symposium*.
27. The best study of the pulp and paper industry in the South is William Clarence Boyd, "New South, New Nature"; also see Oden, "Origins of the Southern Kraft Paper Industry"; Reed, *Crusading for Chemistry*.
28. Maunder, *Voices from the South*, 42; on the importance of the survey in regard to the pulp and paper industry, also see Wakeley, *Biased History*, 165–66.
29. Jeffords, "Trends in Pine and Pulpwood Marketing."
30. Siefkin, "Place of the Pulp and Paper Industry."
31. Jeffords, "Trends in Pine and Pulpwood Marketing," 463.
32. Siefkin, "Place of The Pulp and Paper Industry."
33. R. A. Bonninghausen, "Taking Forestry Services to the Southern Timber Grower and Processor by State Agencies," *Newsletter, Southeastern Section*, Society of American Foresters (1953): 15, in Society of American Foresters Papers, box 102, FHS Archives.
34. J. Walter Myers Jr., "Taking Forestry Services to the Southern Timber Grower and Processor (Industry and Association)," *Newsletter, Southeastern Section*, Society of American Foresters (1953), pp. 18–24, ibid.
35. Jefords, "Trends in Pine and Pulpwood Marketing," 466.
36. S. B. Kinne Jr., "A Look at Short Rotation Pine Management," *Newsletter, Southeastern Section*, Society of American Foresters (1954): 33, in Society of American Foresters Papers, box 102, FHS Archives.
37. "Managing the Small Forest," pamphlet in Georgia Forestry Association Papers, box 51, Hargrett Rare Book and Manuscript Library, UGA, Athens, Ga.
38. Historian Richard White might call this an "organic machine." See White, *Organic Machine*.
39. Kinne, "Look at Short Rotation Pine Management."
40. Memorandum on Brunswick, Ga. Lands, etc. for Messrs. Jenkins and Tift, September 21, 1937, HLS Papers, Hadley Brown Correspondence, TTRS.

41. HLS to John Conway, June 15, 1945, HLS Papers, John Conway Correspondence, TTRS.

42. HLS to Ira Gabrielson, April 26, 1947, HLS Papers, Gabrielson Correspondence, TTRS.

43. J. M. Mallory, "Proceedings of the Sixteenth Annual Meeting of the GFA, May 19–20, 1937," 4, Georgia Forestry Association Papers, box 50, folder 1, Hargrett Rare Book and Manuscript Library, UGA.

44. *LaGrange Daily News*, May 15 and May 23, 1944, in Papers of the National Forest Products Association, box 33, FHS Archives.

45. Chapman, "Do We Want a 'Pulpwood' Economy?" 40–42.

46. The sawtimber interests formed only one group that was critical of pulp and paper. In later years, air and water pollution from coastal paper mills prompted major opposition to the industry. See Cobb, *Selling of the South*, chap. 9; and William Clarence Boyd, "New South, New Nature," chap. 5.

47. Stoddard, *Bobwhite Quail*, 370.

48. On *Dauerwald* and the details of Leopold's trip to Germany, see Meine, *Aldo Leopold*, 351–61; and Newton, *Aldo Leopold's Odyssey*, 292–96.

49. Meine, *Aldo Leopold*, 354–55.

50. Leopold, "Deer and Dauerwald in Germany," 366.

51. From an address titled "Deer and Forestry in Germany," quoted in Meine, *Aldo Leopold*, 360.

52. HLS to Aldo Leopold, June 22, 1946, AL Papers, 9/25/10-1, box 3, Stoddard Correspondence.

53. Stoddard did not worry much with setting an allowable cut on any given tract until he began calculating volume with timber cruisers later in the 1950s. He had worked closely with timber cruisers in assessing real estate values, but did not use them in his forestry work until shortly after World War II. He initially worked out a harvest system that calculated an allowable cut on any given tract of land at about 90 percent of the annual growth increment. That number decreased over the years as his protégé, Leon Neel, refined the system based on the stocking of each individual tract. For a fuller discussion of what came to be known as the Stoddard-Neel approach to forestry, see Neel, Sutter, and Way, *Art of Managing Longleaf*, esp. chap. 4.

54. Herbert L. Stoddard, "The Use of Controlled Fire in Southeastern Game Management" (1939), in Stoddard, Beadel, and Komarek, *Cooperative Quail Study Association*, 190.

55. Ibid.

56. See Stoddard, "Coordinated Forestry, Farming, and Wildlife Programs," 188; and Stoddard, "Relation of Fire," 42.

57. HLS to Aldo Leopold, December 24, 1942, private papers of Leon Neel.

58. HLS to Aldo Leopold, May 16, 1943, ibid.

59. "Memo of Understanding Governing the Sale and Cutting of Poles on Sinkola Land Company and other Plantations of the Thomasville Section," HLS Papers, Contracts, TTRS.

60. These requirements were part of Stoddard's standard contract, of which there are numerous found throughout his papers. For an example, see "Timber Contract, Greenwood Farms and Mitchell Brothers Lumber Company, March 15, 1951," ibid.

61. Aldo Leopold to HLS, September 3, 1943, private papers of Leon Neel.

62. Aldo Leopold to HLS, November 5, 1943, ibid.

63. Aldo Leopold to Daniel Hebard, October 20, 1939, AL Papers, 9/25/10-2, box 4, file 10.

64. Frank Heyward Jr. to HLS, June 23, 1938, HLS Papers, Fire Correspondence, TTRS.

65. HLS to Frank Heyward Jr., June 27, 1938, ibid.

66. Stoddard, "Use of Controlled Fire," 189.

67. Frank Heyward Jr. to HLS, May 24, 1939, HLS Papers, Fire Correspondence, TTRS.

68. Ibid.

69. I should note that Heyward's early training was at the Southern Forest Experiment Station, and besides the implications about uncontrolled fire, he personally had "failed to find that any of [Stoddard's] statements are contrary to my own opinions" (ibid.). Heyward had published several articles favorable of fire while employed by the Forest Service. See, for instance, Heyward, "Relation of Fire to Stand Composition."

70. HLS to Frank Heyward Jr., May 26, 1939, HLS Papers, Fire Correspondence, TTRS.

71. Ibid.

72. Stoddard, "Coordinated Forestry, Farming, and Wildlife Programs," 187–88.

73. Stoddard, "Relation of Fire," 42.

74. HLS to Richard Parks, April 10, 1950, HLS Papers, Richard Parks Correspondence, TTRS.

75. See Neel, Sutter, and Way, *Art of Managing Longleaf.*

76. *Tall Timbers Fire Ecology Conference, Proceedings*, 1 (1962): 156–57.

Chapter 7. Bringing Agrarian Science to the Public

1. HLS to Ovid Butler, September 26, 1931, HLS Papers, Fire Correspondence, TTRS.

2. HLS to Ovid Butler, December 5, 1931; Ovid Butler to HLS, October 9, 1931, ibid.

3. Hays, *Beauty, Health, and Permanence*; Rome, *Bulldozer in the Countryside*; Worster, *Nature's Economy*; Rothman, *The Greening of a Nation?*

4. Komarek, "History of Prescribed Fire," 8.

5. On the importance of the Sunday morning gatherings in the founding of Tall Timbers, see Neel, Sutter, and Way, *Art of Managing Longleaf*, 116–19; Komarek, *Quest for Ecological Understanding*, 21; and Rubanowice, *Sense of Place*, 258–65.

6. Komarek, *Quest for Ecological Understanding*, 9.

7. Ibid., 10.

8. On the Komareks at Greenwood, see Komarek, "Role of Tall Timbers Research Station"; and Komarek, "History of Prescribed Fire." Ed Komarek was also heavily involved in the development of hybrid corn varieties for the South, working closely with crop geneticists at the Georgia Coastal Plain Experiment Station in Tifton. With Whitney's support, he started the Greenwood Seed Company in the mid-1950s and sold the popular Dixie-18 hybrid for several years. See above publications, and Neel, Sutter, and Way, *Art of Managing Longleaf*, 119–20.

9. "Christmas Greetings 1958," HLS Papers, TTRS.

10. Crawford, *Great Effort*, 40.

11. See Stoddard, *Memoirs of a Naturalist*, 283.

12. Earle R. Greene et al., *Birds of Georgia*; Burleigh, *Birds of Georgia*. Burleigh dedicated the book to Stoddard. On Odum's life and career, see Craige, *Eugene Odum*; Golley, *History of the Ecosystem Concept*; Van Sant, "Representing Nature, Reordering Society."

13. Most of the checklist correspondence is in the Eugene Odum Papers, Hargrett Rare Book and Manuscript Library, UGA, University Archives, Accession Number 97-045, boxes 51–52.

14. Stoddard was especially involved in the work of Robert Norris, a master's student of Odum's who would come back to work at Tall Timbers after completing his doctoral work at Stanford. See Norris, *Distribution and Populations of Summer Birds*.

15. Barrow, *Passion for Birds*, chap. 6.

16. A host of publications resulted from the tower project. For overviews see, Stoddard, "Bird Casualties"; Stoddard and Norris, "Bird Casualties"; Crawford, "Bird Casualties"; Odum and his students produced almost twenty publications from the tower birds. For a sampling, see Odum, "Lipid Deposition"; Odum, Connell, and Stoddard, "Flight Energy and Estimated Flight Ranges"; for a complete bibliography, see Komarek, *Quest for Ecological Understanding*, 103–5; also see Robert L. Crawford's interesting history of the tower study, *The Great Effort*.

17. Two exceptions include Komarek, "Mammal Relationships to Upland Game"; and Komarek, "Progress Report on Southeastern Mammals."

18. There is a large literature on the growth of "big science." On biology, see Allen, *Life Science*; and Rainger, Benson, Maienschein, *Expansion of American Biology*, esp. Rainger's introduction.

19. HLS to Odum, January 10, 1960, HLS Papers, Odum Correspondence, TTRS.

20. Cypert, "Effects of Fires."

21. Herbert L. Stoddard, "Drought" manuscript, private papers of Leon Neel.

22. Odum to Stoddard, November 10, 1960, HLS Papers, Odum Correspondence, TTRS.

23. "Comments on Stoddard Manuscript, by E. P. Odum," ibid.

24. Odum's remarks here should not diminish his respect for the natural history tradition, or for Stoddard's past work and his knowledge of the woods. In one remembrance, he called Stoddard "one of the best all-around naturalists that I have known personally. I would not swap my days in the field with him for a million dollars" (Odum, "Southeastern Region," 2).

25. Komarek, *Quest for Ecological Understanding*, 35.

26. Ibid., 75–76.

27. Ibid., 35.

28. Ibid., 36.

29. Ibid., viii.

30. Ibid.

31. Ibid., 2.

32. E. V. Komarek, "Fire Ecology," in *Tall Timbers Fire Ecology Conference, Proceedings*, 1 (1962): 95.

33. Pyne, *Tending Fire*, 11; on fire science and the Forest Service, also see Pyne, *Fire in America*, 467–96.

34. *Tall Timbers Fire Ecology Conference, Proceedings*, 1 (1962): v.

35. Henry Beadel, "Fire Impressions," in ibid., 4–5.

36. R. A. Bonninghausen, "The Florida Forest Service and Controlled Burning," in ibid., 43.

37. Thomas Lotti, "The Use of Prescribed Fire in the Silviculture of Loblolly Pine," in ibid., 109.

38. Leopold et al., "Wildlife Management in the National Parks."

39. *Tall Timbers Fire Ecology Conference, Proceedings*, 2 (1963): xi–xv.

40. On the changes in fire management in the National Park Service and the U.S. Forest Service policy, see Pyne, *Fire in America*, 300–302, 490–96.

41. *Tall Timbers Research Station Fire Ecology Plots*.

42. Neel, Sutter, and Way, *Art of Managing Longleaf*, 128.

43. Stoddard, *Bobwhite Quail*, 193; on the misidentification of the thief ant, see Stoddard, Beadel, and Komarek, *Cooperative Quail Study Association*, 37.

44. Bernard V. Travis, "The Present Status of Methods of Control for the Fire

Ant," in Stoddard, Beadel, Komarek, *Cooperative Quail Study Association*, 225–31.

45. E. V. Komarek Sr., "Environmental Management," in *Tall Timbers Conference on Ecological Animal Control by Habitat Management, Proceedings*, 8.

46. "Seventh Annual Report of the Cooperative Quail Study Association," in Stoddard, Beadel, Komarek, *Cooperative Quail Study Association*, 172.

47. Blu Buhs, *Fire Ant Wars*, 40. Also see Daniel, "Rogue Bureaucracy."

48. Blu Buhs, *Fire Ant Wars*, 73.

49. See the extensive correspondence with Peters and many others on fire ants in the HLS Papers, Fire Ant Correspondence, TTRS.

50. HLS, "Christmas Greetings, 1958," HLS Papers, TTRS.

51. Leon Neel, interview with author, August 24, 2004.

52. HLS, "Christmas Greetings, 1958"; "Christmas Greetings, 1961," HLS Papers, TTRS.

53. The show was filmed live, and, unfortunately, no recordings or transcripts exist. For reference to Peters's appearance, see Blu Buhs, *Fire Ant Wars*, 130.

54. Komarek, "Environmental Management," 10.

55. Ibid., 7.

56. See Dunlap, *DDT*.

57. Komarek, "Environmental Management," 4–5.

58. Ibid., 5.

Conclusion

1. See Gromme, "In Memoriam."

2. Herbert Stoddard to Eugene Odum, January 26, 1952, HLS Papers, Eugene Odum Correspondence, TTRS.

3. Aldo Leopold to Herbert Stoddard, February 11, 1930. AL Papers, 9/25/10-1, box 3—Stoddard Correspondence.

4. Jose, Jokela, and Miller, *Longleaf Pine Ecosystem*, 4.

5. The best source on current restoration efforts is Earley, *Looking for Longleaf*; also see Neel, Sutter, and Way, *Art of Managing Longleaf*.

BIBLIOGRAPHY

Primary Sources

MANUSCRIPT COLLECTIONS

Chicago Field Museum Archives, Chicago, Illinois
Administrative Papers
Harris Loan Center Papers

Emory University, Robert W. Woodruff Library
Robert W. Woodruff Papers

Forest History Society Archives, Durham, North Carolina
American Forestry Association
National Lumber Manufacturers Association
National Forest Products Association
Society of American Foresters
American Pulpwood Association
Oral History Collection

Joseph W. Jones Ecological Research Station at Ichauway, Newton, Georgia
Ichauway Documentary Project

Library of Congress Manuscripts
W. L. McAtee Papers

Milwaukee Public Museum, Milwaukee, Wisconsin
Administrative Papers
Owen Gromme Papers

National Archives Records Administration, College Park, Maryland
Records of the U.S. Bureau of Biological Survey
Records of the U.S. Forestry Service

Parks Library, Iowa State University
Paul Errington Papers

Private Papers of Leon Neel, Thomasville, Georgia

Tall Timbers Research Station Archives, Tallahassee, Florida
Henry Beadel Papers
Edwin V. Komarek Papers
Herbert L. Stoddard Papers
Oral History Collection

Thomas County Historical Society, Thomasville, Georgia
Brandon Collection
H. W. Hopkins Collection
Moore Collection

Thomasville Genealogical, History, and Fine Arts Library, Thomasville, Georgia
Elizabeth Hopkins Collection

University of Georgia, Hargrett Rare Book and Manuscript Library
Eugene Odum Papers
Georgia Forestry Association Papers
Southern Pulpwood Conservation Association Papers
Thomas Jones Papers

University of Wisconsin–Madison, Steenbock Memorial Library
Aldo Leopold Papers

Interviews by Author

Jimmy Atkinson
Jerry Franklin
Angus Gholson
Jimmy Greene
Leon Neel

Newspapers and Periodicals

Atlanta Constitution
Atlanta Journal
American Forests
Georgia Farmer
LaGrange Daily News
New York Times
Progressive Farmer
Tallahassee Daily Democrat
Tallahassee Semi-Weekly True Democrat
Tallahassee Weekly True Democrat
Tallahassee Weekly Floridian
Thomasville Daily Times-Enterprise

Thomasville Times
Thomasville Times Enterprise
Thomasville Weekly Times-Enterprise and South Georgia Progress
Weekly Floridian

PUBLISHED DOCUMENTS

Acts and Resolutions of the General Assembly of the State of Georgia, 1876.

Akeley, Mary Jobe. *The Wilderness Lives Again: Carl Akeley and the Great Adventure*. New York: Dodd, Mead and Company, 1940.

Ashe, W. W. "Forest Fires: Their Destructive Work, Causes and Prevention." Bulletin No. 7, North Carolina Geological Survey. Raleigh, 1895.

———. "The Forests, Forest Lands, and Forest Products of Eastern North Carolina." Bulletin No. 5, North Carolina Geological Survey. Raleigh, 1894.

———. "The Long-Leaf Pine and Its Struggle for Existence." *Journal of the Elisha Mitchell Scientific Society* 11 (1894): 1–16.

Babcock, Haveliah. *My Health Is Better in November: Thirty-Five Stories of Hunting and Fishing in the South*. Columbia: University of South Carolina Press, 1947.

Barbour, George M. *Florida for Tourists, Invalids, and Settlers*. 1882. Reprint, Gainesville: University of Florida Press, 1964.

Beard, George M. *American Nervousness: Its Causes and Consequences*. New York: G. P. Putnam's Sons, 1881.

Bill, Ledyard. *A Winter in Florida*. New York: Wood and Holbrook, 1869.

Brinton, Daniel G. *A Guide-Book of Florida and the South, for Tourists, Invalids and Emigrants*. 1869. Reprint, Gainesville: University Press of Florida, 1978.

Brockway, Dale G., et al. *Restoration of Longleaf Pine Ecosystems*. General Technical Report SRS-83. Asheville, N.C.: Southern Research Station, U.S. Forest Service, U.S. Department of Agriculture, 2005.

———. *Uneven-Aged Management of Longleaf Pine Forests: A Scientist and Manager Dialogue*. General Technical Report-78. Asheville, N.C.: Southern Research Station, U.S. Forest Service, U.S. Department of Agriculture, 2005.

Brown, Titus, and James Hadley. *African American Life on the Southern Hunting Plantation*. Charleston, S.C.: Arcadia Publishing, 2000.

Bull, Henry, and R. R. Reynolds. "Management of Loblolly Pine: Further Study Needed." *Journal of Forestry* 41, no. 10 (1943): 722–29.

Burleigh, Thomas D. *Birds of Georgia*. Norman: University of Oklahoma Press, 1958.

Chapin, George M. *Florida Past, Present and Future*. 2 vols. Chicago, 1914.

Chapman, H. H. "Common Sense Needed." *Journal of Forestry* 41, no. 10 (1943): 722–29.

———. "Do We Want a 'Pulpwood' Economy for Our Southern Forests?" *American Forests* 60, no. 1 (1954): 22–23, 40–44.

———. "Factors Determining Natural Reproduction of Longleaf Pine on Cut-over Lands in La Salle Parish, La." *Yale University School of Forestry Bulletin* 16 (1926): 1–44.

———. "Is the Longleaf Type a Climax?" *Ecology* 13, no. 4 (1932): 328–34.

———. "Management of Loblolly Pine in the Pine-Hardwood Region in Arkansas and in Louisiana West of the Mississippi River." *Yale University School of Forestry Bulletin* 49 (1942): 1–150.

———. "Some Further Relations of Fire to Longleaf Pine." *Journal of Forestry* 30 (1932): 602–4.

Chapman, H. H., and R. Bulchis. "Increased Growth of Longleaf Pine Seed Trees at Urania, La., after Release Cutting." *Journal of Forestry* 38 (1940): 722–26.

Cowles, Henry. "Ecological Relations of Vegetation on the Sand Dunes of Lake Michigan." *Botanical Gazette* 27 (1899): 361–91.

Clements, Frederic E. "The Relict Method in Dynamic Ecology." *Journal of Ecology* 22 (1934): 39–68.

Crawford, Robert L. "Bird Casualties at a Leon County, Florida TV Tower: A 25-Year Migration Study." *Bulletin*, no. 22 (July 1981): 1–30.

Cromwell, Benjamin M. "The Influence of Trees on Health." *Biannual Report*. Georgia Board of Health, 1875.

Cypert, Eugene. "The Effects of Fires in the Okefenokee Swamp in 1954 and 1955." *American Midland Naturalist* 66, no. 2 (1961): 485–503.

Demmon, E. L. "The Silvicultural Aspects of the Forest-Fire Problem in the Longleaf Pine Region." *Journal of Forestry* 33 (March 1935): 325.

———. "Twenty Years of Forest Research in the Lower South, 1921–1941." *Journal of Forestry* 10, no. 1 (1942): 33–36.

Dyer, C. Dorsey, and C. Nelson Brightwell. "Prescribed Burning in Slash and Longleaf Pine Forests of Georgia." Athens: Georgia Agricultural Extension Service, 1955.

Eldredge, Inman. "Administrative Problems in Fire Control in the Longleaf-Slash Pine Region of the South." *Journal of Forestry* 33 (March 1935): 344.

Elton, Charles S. *Animal Ecology*. 1927. Reprint, Chicago: University of Chicago Press, 2001.

———. *The Ecology of Invasions of Plants and Animals*. 1958. Reprint, Chicago: University of Chicago Press, 2000.

Errington, Paul L. "The Northern Bobwhite: Environmental Factors Influencing Its Status." PhD diss., University of Wisconsin–Madison, 1932.

———. "Over-Populations and Predation: A Research Field of Singular Promise." *The Condor* 37, no. 5 (1935): 230–32.

———. "Predator and Vertebrate Populations." *Quarterly Review of Biology* 21, no. 2 (1946): 144–77.

———. "Vulnerability of Bob-White Populations to Predation." *Ecology* 15, no. 2 (1934): 110–27.

Errington, Paul L., and H. L. Stoddard. "Modifications in Predation Theory Suggested by Ecological Studies of the Bobwhite Quail." In *Transactions of the Third North American Wildlife Conference*, 736–40. Washington, D.C.: American Wildlife Institute, 1938.

Farrar, Robert M., ed. *Proceedings of the Symposium on the Management of Longleaf Pine*. General Technical Report SO-75. New Orleans: Southern Forest Experiment Station, U.S. Forest Service, U.S. Department of Agriculture, 1990.

Field, Henry M. *Bright Skies and Dark Shadows*. New York: Charles Scribner's Sons, 1890.

The Fire Forest: Longleaf Pine–Wiregrass Ecosystem. Covington: Georgia Wildlife Press, Georgia Wildlife Federation, 2001.

Fisher, Albert K. *The Hawks and Owls of the United States in Their Relation to Agriculture*. Bulletin no. 3. Washington, D.C.: Division of Economic Ornithology and Mammalogy, U.S. Department of Agriculture, 1893.

"Forest Facts for Georgia." Georgia Agriculture and Industrial Development Board, 1945.

Gatewood, Steve, et al. *A Comprehensive Study of a Portion of the Red Hills Region of Georgia*. Thomasville: Thomas College Press, 1994.

Greeley, W. B. "The Relation of Geography to Timber Supply." *Economic Geography* 1 (March 1925): 7.

Greene, Earle R., William W. Griffin, Eugene P. Odum, Herbert L. Stoddard, and Ivan R. Tomkins. *Birds of Georgia: A Preliminary Check-list and Bibliography of Georgia Ornithology*. Occasional Publication no. 2, Georgia Ornithological Society. Athens: University of Georgia Press, 1945.

Greene, S. W. "The Forests That Fire Made." *American Forests*, October 1931, 583–84.

Gromme, Owen J. "In Memoriam: Herbert Lee Stoddard." *The Auk* 90 (October 1973): 870–76

Hallock, Charles. *Camp Life in Florida; A Handbook for Sportsmen and Settlers*. New York: Forest and Stream Publishing Company, 1876.

Hardtner, Henry E. "Henry E. Hardtner, Urania Lumber Company, Urania, La., Discusses Reforestation and Controlled Burnings." *Lumber Trade Journal* 74, no. 10 (1918): 35–36.

———. "A Tale of a Root—A Root of a Tale or, Root Hog or Die." *Journal of Forestry* 33 (March 1935): 351–60.

Harper, Roland M. *Economic Botany of Alabama*. Monograph no. 8. University, Ala.: Geological Survey of Alabama, 1913.

———. "Fire and Forests." *American Botany* 46 (1940): 5–7.

———. "The Forest Resources of Alabama." *American Forests* 19 (1913): 657–70.

———. "Geography and Vegetation of Northern Florida." *Florida State Geological Survey, Sixth Annual Report*. E. H. Sellards, ed. Tallahassee: State Geological Survey, 1914.

———. "History of Soil Investigation in Florida." *Florida Geological Survey, Seventeenth Annual Report, 1924–1925*. Tallahassee: State Geological Survey, 1926.

———. "The Natural Resources of an Area in Central Florida." *Florida Geological Survey, Seventh Annual Report*. Tallahassee: Florida Geological Survey, 1915.

———. "The Relation of Climax Vegetation to Islands and Peninsulas." *Bulletin of the Torrey Botanical Club* 38 (1911): 515–25.

Heinsohn, Lillian Britt. *Southern Plantation: The Story of Labrah*. New York: Bonanza Books, 1962.

Heyward, Frank. "The Relation of Fire to Stand Composition of Longleaf Forests." *Ecology* 20 (April 1939): 287–304.

Holling, C. S. *Adaptive Environmental Assessment and Management*. New York: John Wiley and Sons, 1978.

Hopkins, Thomas N. "The Pine Forests of Southern Georgia, Its Climate and Adaptability to the Consumptive, *Thomas County, Georgia*." Thomasville, Ga.: Thomas County Immigration Society, 1877.

Hornaday, William T. "Common Faults in the Mounting of Quadrupeds." *Annual Report of the Society of American Taxidermists* 3 (1883): 67–71.

Howell, P. N. "My Experience with Fire in Longleaf Pine." *American Forests*, March, 1932.

Humphrey, George M., and Dr. Shepard Krech, eds. *The Georgia–Florida Field Trial Club, 1916–1948*. New York: Scribner Press, 1948.

Jeffords, A. I., Jr. "Trends in Pine and Pulpwood Marketing in the South." *Journal of Forestry* 54, no. 7 (1956): 463–66.

Jose, Shibu, Eric J. Jokela, and Deborah L. Miller, eds. *The Longleaf Pine Ecosystem: Ecology, Silviculture, and Restoration*. New York: Springer, 2006.

Kellar, Herbert Anthony, ed. *Solon Robinson, Pioneer and Agriculturist, 1803–1880*. New York: Da Capo Press, 1968.

Komarek, Edwin V. "History of Prescribed Fire and Controlled Burning in Wildlife Management in the South." In *Prescribed Fire and Wildlife in Southern Forests*, edited by Gene W. Wood, 1–14. Georgetown, S.C.: Belle W. Baruch Forest Science Institute of Clemson University, 1981.

———. "Mammal Relationships to Upland Game and Other Wildlife." *Second North American Wildlife Conference*, February 1937, 561–69.

———. "A Progress Report on Southeastern Mammals." *Journal of Mammalogy* 20, no. 3 (1939): 292–99.

———. *A Quest for Ecological Understanding: The Secretary's Review, March 15, 1958—June 30, 1975*. Miscellaneous Publication no. 5. Tallahassee: Tall Timbers Research Station, 1977.

———. "The Role of Tall Timbers Research Station in the Development of the Study of Fire Ecology." In *Symposium on the Environmental Consequences of Fire and Fuel Management in Mediterranean Ecosystems*, edited by H. A. Mooney and C. E. Conrad, 488–98. Washington, D.C.: USDA Forest Service, 1977.

Leopold, Aldo. "Deer and Dauerwald in Germany, Part I: History." *Journal of Forestry* 34, no. 4 (1936): 366–75.

———. *For the Health of the Land: Previously Unpublished Essays and Other Writings*. Edited by J. Baird Callicott and Eric T. Freyfogle. Washington, D.C.: Island Press, 1999.

———. *Game Management*. New York: Charles Scribner's Sons, 1933.

———. *The River of the Mother of God*. Edited by Susan L. Flader and J. Baird Callicott. Madison: University of Wisconsin Press, 1991.

———. *A Sand County Almanac; and Sketches Here and There*. Commemorative ed. New York: Oxford University Press, 1989.

Leopold, A. S., S. A. Cain, C. M. Cottam, I. N. Gabrielson, and T. L. Kimball. "Wildlife Management in the National Parks." March 4, 1963. Available at http://www.cr.nps.gov/history/online_books/leopold/leopold.htm.

Long, Ellen Call. "Forest Fires in the Southern Pines." *Forest Leaves* 2 (1889): 94.

Maunder, Elwood R., ed. *Voices from the South: Recollections of Four Foresters*. Santa Cruz, Calif.: Forest History Society, 1977.

McAtee, W. L. "Effectiveness in Nature of the So-Called Protective Adaptations in the Animal Kingdom, Chiefly as Illustrated by the Food Habits of Nearctic Birds." *Smithsonian Misc. Coll.* 85, no. 7 (1932): 1–201.

McIntyre, R. K., S. B. Jack, R. J. Mitchell, J. K. Hiers, W. L. Neel. *Multiple Value Management: The Stoddard-Neel Approach to Ecological Forestry in Longleaf Pine Grasslands*. N.p.: Joseph W. Jones Ecological Research Center, 2008.

Mitchell, Robert J., J. Kevin Hiers, Joseph J. O'Brien, S. B. Jack, and R. T. Engstrom. "Silviculture That Sustains: The Nexus between Silviculture, Frequent Prescribed Fire, and Conservation Biodiversity in Longleaf Pine Forests of the Southeastern United States." *Canadian Journal of Forest Resources* 36 (2006): 2724–36.

Mitchell, Robert J., J. Kevin Hiers, Joseph O'Brien, and Gregory Starr. "Ecological Forestry in the Southeast: Understanding the Ecology of Fuels." *Journal of Forestry* 107, no. 8 (2009): 391–97.

Norris, Robert. *Distribution and Populations of Summer Birds in Southwestern Georgia*. Occasional Publication no. 3, Georgia Ornithological Society. Athens: University of Georgia Press, 1951.

Odum, Eugene P. "Lipid Deposition in Nocturnal Migrant Birds." *Proceedings of the 12th International Ornithological Congress* (1960): 563–76.

———. "The Southeastern Region: A Biodiversity Haven for Naturalists and Ecologists," *Southeastern Naturalist* 1, no. 1 (2002): 2.

Odum, Eugene P., C. E. Connell, and Herbert L. Stoddard. "Flight Energy and Estimated Flight Ranges of Some Migratory Birds." *Auk* 78 (1961): 515–27.

Pearson, G. A. "Intensive Forestry Needed." *Journal of Forestry* 41, no. 10 (1943): 722–29.

Peterson, Roger Tory, and James Fisher. *Wild America*. Boston: Houghton Mifflin, 1955.

Schantz, Orpheus Moyer. "Indiana's Unrivaled Sand-Dunes: A National Park Opportunity." *National Geographic* 35 (May 1919): 430–41.

Schorger, A. W. "In Memoriam: Paul Lester Errington." *The Auk* 83 (January 1966): 52–65.

Schultz, Robert P. *Loblolly Pine: The Ecology and Culture of the Loblolly Pine*. Washington, D.C.: U.S. Department of Agriculture Forest Service, 1997.

Shea, John. "Getting at the Roots of Man-Caused Forest Fires: A Case Study of a Southern Forest Area." Washington, D.C.: U.S. Forest Service, ca. 1939.

———. "Man-Caused Forest Fires: The Psychologist Makes a Diagnosis." Washington, D.C.: U.S. Forest Service, 1939.

———. "Our Pappies Burned the Woods," *American Forestry* 46 (April 1940): 159–74.

Shrosbree, George. "A Collecting Expedition to Bonaventure Island, Quebec." *Yearbook of the Public Museum of the City of Milwaukee* 2 (1922): 22.

Siefkin, Gordon. "The Place of the Pulp and Paper Industry in the Georgia Economy." Atlanta: Emory University, School of Business Administration, December 1957.

Stoddard, Herbert L. "Bird Banding in Milwaukee and Vicinity." *Yearbook of the Public Museum of the City of Milwaukee* 3 (1923): 117.

———. "Bird Casualties at a Leon County, Florida TV Tower, 1955–1961." *Bulletin of Tall Timbers Research Station*, no. 1, 1–94. Tallahassee: Tall Timber Research Station, 1962.

———. *The Bobwhite Quail: Its Habits, Preservation, and Increase*. New York: Charles Scribner and Sons, 1931.

———. "Coordinated Forestry, Farming, and Wildlife Programs for Family Sized Farms of the Coastal Plain of the Deep Southeast." *Proceedings, Society of American Foresters Meeting* (October 15–17, 1956), 186–89. Washington, D.C.: Society of American Foresters, 1957.

———. "Local Bird Notes." *Yearbook of the Public Museum of the City of Milwaukee* 3 (1923): 128–29.

———. *Memoirs of a Naturalist*. Norman: University of Oklahoma Press, 1969.

———. "Preliminary Report on Cooperative Quail Investigation: 1924." Washington, D.C.: Committee of the Quail Study Fund for Southern Georgia and Northern Florida, U.S. Biological Survey, 1924.

———. "The Relation of Fire to the Game of the Forest." In *Special Problems in Southern Forest Management, Proceedings of the Sixth Annual Forestry Symposium*, 36–45. Baton Rouge: School of Forestry, Louisiana State University, April 4–5, 1957.

———. "Report on Cooperative Quail Investigation: 1925–1926: With Preliminary Recommendations for the Development of Quail Preserves." Washington, D.C.: Committee Representing the Quail Study Fund for Southern Georgia and Northern Florida, U.S. Biological Survey, 1926.

———. "Stuffed Birds." *Yearbook of the Public Museum of the City of Milwaukee* 2 (1922): 182–85.

Stoddard, Herbert L., Henry Beadel, and Edwin V. Komarek. *The Cooperative Quail Study Association, May 1, 1931–May 1, 1943*. Miscellaneous Publication, no. 1. Tallahassee: Tall Timbers Plantation, 1961.

Stoddard, Herbert L., and R. A. Norris, "Bird Casualties at a Leon County, Florida TV Tower: An Eleven-Year Study, *Bulletin of Tall Timbers Research Station*, no. 8, 1–104. Tallahassee: Tall Timbers Research Station, 1967.

Tall Timbers Conference on Ecological Animal Control by Habitat Management, Proceedings. Tallahassee: Tall Timbers Research Station, 1969.

Tall Timbers Fire Ecology Conference, Proceedings, 15 vols. Tallahassee: Tall Timbers Research Station, 1962–74.

Tall Timbers Research Station Fire Ecology Plots, Bulletin no. 2. Tallahassee: Tall Timbers Research Station, 1962.

Torrey, Bradford. *A Florida Sketch-Book*. Boston: Houghton Mifflin Company, 1894.

Triplett, John. *Thomasville, Georgia: The Health Resort*. Thomasville, Ga.: John Triplett, 1891.

U.S. Bureau of the Census. *Tenth Census of Agriculture*. Washington, D.C.: Government Printing Office, 1880.

———. *Fifteenth Census of Agriculture*. Washington, D.C.: Government Printing Office, 1930.

U.S. Department of Agriculture, National Resources Conservation Service. *Soil Taxonomy: A Basic System of Soil Classification for Making and Interpreting Soil Surveys*. Agricultural Handbook, no. 436. Washington, D.C.: U.S. Government Printing Office, 1999.

U.S. Department of Agriculture, Soil Conservation Service. *Soil Survey of Brooks and Thomas Counties, Georgia*. Athens: University of Georgia, College of Agriculture, and Agriculture Experiment Stations, May 1979.

Vance, Rupert B. *Human Geography of the South: A Study in Regional Resources and Human Adequacy*. Chapel Hill: University of North Carolina Press, 1932.

Wahlenberg, W. G., ed. *A Guide to Loblolly and Slash Pine Plantation Management in Southeastern USA*. Macon: Georgia Forest Research Council, 1965.

———. *Loblolly Pine: Its Use, Ecology, Regeneration, Protection, Growth, and Management*. Durham, N.C.: School of Forestry, Duke University, 1960.

———. *Longleaf Pine: Its Use, Ecology, Regeneration, Protection, Growth, and Management*. Washington, D.C.: Charles Lathrop Pack Forestry Foundation in cooperation with the U.S. Forest Service, U.S. Department of Agriculture, 1946.

Wakeley, Philip C. "The Adolescence of Forestry Research in the South." *Journal of Forest History* 22, no. 3 (1978): 136–45.

———. *Biased History of the Southern Forest Experiment Station through Fiscal Year 1933*. New Orleans: Southern Forest Experiment Station, 1979.

Wing, Leonard William. "Naturalize the Forest for Wildlife." *American Forests* 42 (June 1936): 260–61, 293.

Wooten, H. H. *The Land Utilization Program, 1934 to 1964*. Washington, D.C.: Government Printing Office, 1965.

Secondary Sources

Ackerman, Joe A., Jr. *Florida Cowman: A History of Florida Cattle Raising*. Kissimmee: Florida Cattlemen's Association, 1976.

Aiken, Charles. *The Cotton Plantation South since the Civil War*. Baltimore: Johns Hopkins University Press, 1998.

Allen, Garland. *Life Science in the Twentieth Century*. Cambridge: Cambridge University Press, 1975.

Armitage, Kevin C. *The Nature Study Movement: The Forgotten Popularizer of America's Conservation Movement*. Lawrence: University Press of Kansas, 2009.

Aron, Cindy. *Working at Play: A History of Vacations in the United States*. New York: Oxford University Press, 1999.

Arsenault, Raymond. "The End of the Long Hot Summer: The Air Conditioner and Southern Culture." *Journal of Southern History* 50 (November 1984): 597–628.

Ayers, Edward L. *The Promise of the New South: Life after Reconstruction*. New York: Oxford University Press, 1992.

Balogh, Brian. "Scientific Forestry and the Roots of the Modern American State: Gifford Pinchot's Path to Progressive Reform." *Environmental History* 7, no. 2 (2002): 198–225.

Baptist, Edward E. *Creating an Old South: Middle Florida's Plantation Frontier before the Civil War*. Chapel Hill: University of North Carolina Press, 2002.

Barrow, Mark V., Jr. *Nature's Ghosts: Confronting Extinction from the Age of Jefferson to the Age of Ecology*. Chicago: University of Chicago Press, 2009.

———. *A Passion for Birds: American Ornithology after Audubon*. Princeton: Princeton University Press, 1998.

———. "Science, Sentiment, and the Specter of Extinction: Reconsidering Birds of Prey during America's Interwar Years." *Environmental History* 7, no. 1 (2002): 69–98.

———. "The Specimen Dealer: Entrepreneurial Natural History in America's Gilded Age." *Journal of the History of Biology* 33, no. 3 (2000): 493–534.

Bartley, Numan V. *The New South, 1945–1980: The Story of the South's Modernization*. Baton Rouge: Louisiana State University Press, 1995.

Bates, Barbara. *Bargaining for Life: A Social History of Tuberculosis, 1876–1938*. Philadelphia: University of Pennsylvania Press, 1992.

Bauer, Robin Theresa. "At Home among the Red Hills: The African American Tenant Farm Community on Tall Timbers Plantation." MA thesis, Florida State University, 2005.

Bederman, Gail. *Manliness and Civilization: A Cultural History of Gender and Race in the United States, 1880–1917*. Chicago: University of Chicago Press, 1995.

Beeman, Randal S., and James A. Pritchard. *A Green and Permanent Land: Ecology and Agriculture in the Twentieth Century*. Lawrence: University Press of Kansas, 2001.

Belanger, Dian Olson. *Managing American Wildlife: A History of the International Association of Fish and Wildlife Agencies*. Amherst: University of Massachusetts Press, 1988.

Bermingham, Ann. *Landscape and Ideology: The English Rustic Tradition, 1740–1860*. Berkeley: University of California Press, 1986.

Blight, David W. *Race and Reunion: The Civil War in American Memory*. Cambridge, Mass.: Harvard University Press, 2001.

Bloch, Marc. *French Rural History: An Essay on Its Basic Characteristics*. Berkeley: University of California Press, 1966.

Blum, Jerome. *The End of the Old Order in Rural Europe*. Princeton: Princeton University Press, 1978.

Bodry-Sanders, Penelope. *Carl Akeley: Africa's Collector, Africa's Savior*. New York: Paragon House, 1991.

Boyd, William Clarence. "The Forest Is the Future? Industrial Forestry and the Southern Pulp and Paper Complex." In *The Second Wave: Southern Industrialization from the 1940s to the 1970s*, edited by Philip Scranton, 168–218. Athens: University of Georgia Press, 2001.

———. "New South, New Nature: Regional Industrialization and Environmental Change in the Post–New Deal American South." PhD diss., University of California, Berkeley, 2002.

Braden, Susan R. *The Architecture of Leisure: The Florida Resort Hotels of Henry Flagler and Henry Plant*. Gainesville: University of Florida Press, 2002.

Brown, Margaret Lynn. *The Wild East: A Biography of the Great Smoky Mountains*. Gainesville: University Press of Florida, 2000.

Brown, Titus. "The African-American Middle Class in Thomas County, Georgia, in the Age of Booker T. Washington." *Journal of South Georgia History*, Fall 2000, 55–76.

Brubaker, Harry F. "Land Classification, Ownership and Use in Leon County, Florida." PhD diss., University of Michigan, 1956.

Brueckheimer, William R. *Leon County Hunting Plantations: An Historical and Architectural Survey*. Tallahassee: Historic Tallahassee Preservation Board, 1988.

Buhs, Joshua Blu. *The Fire Ant Wars: Nature, Science, and Public Policy in Twentieth-Century America*. Chicago: University of Chicago Press, 2004.

Cameron, Jenks. *The Bureau of Biological Survey: Its History, Activities, and Organization*. Baltimore: Johns Hopkins University Press, 1929.

Carney, Judith. *Black Rice: The African Origins of Rice Cultivation in the Americas*. Cambridge, Mass.: Harvard University Press, 2001.

Cioc, Marc. *The Game of Conservation: International Treaties to Protect the World's Migratory Animals*. Athens: Ohio University Press, 2009.

Cobb, James C. *Away Down South: A History of Southern Identity*. New York: Oxford University Press, 2005.

———. "Beyond Planters and Industrialists: A New Perspective on the New South." *The Journal of Southern History* 54, no. 1 (1988): 45–68.

———. *Industrialization and Southern Society, 1877–1984*. Lexington: University of Kentucky Press, 1984.

———. *The Most Southern Place on Earth: The Mississippi Delta and the Roots of Regional Identity*. New York: Oxford University Press, 1992.

———. *The Selling of the South: The Southern Crusade for Industrial Development, 1936–1990*. 2nd ed. Urbana: University of Illinois Press, 1993.

Coleman, Jon T. *Vicious: Wolves and Men in America*. New Haven: Yale University Press, 2004.

Cowdrey, Albert E. *This Land, This South: An Environmental History*. Rev. ed. Lexington: University Press of Kentucky, 1996.

Cox, Thomas R. *This Well-Wooded Land: Americans and their Forests from Colonial Times to the Present*. Lincoln: University of Nebraska, 1985.

Craige, Betty Jean. *Eugene Odum: Ecosystem Ecologist and Environmentalist*. Athens: University of Georgia Press, 2002.

Craven, Avery O. *Soil Exhaustion as a Factor in the Agricultural History of Virginia and Maryland, 1606–1860*. Urbana: University of Illinois Press, 1925.

Crawford, Robert L. *The Great Effort: Herbert L. Stoddard and the WCTV Tower Study*. Misc. Pub. no. 14. Tallahassee: Tall Timbers Research Station, 2004.

Croker, Robert A. *Pioneer Ecologist: The Life and Work of Victor Ernest Shelford, 1877–1968*. Washington, D.C.: Smithsonian Institution Press.

Cronon, William, *Changes in the Land: Indians, Colonists, and the Ecology of New England*. New York: Hill and Wang, 1983.

———. *Nature's Metropolis: Chicago and the Great West*. New York: W. W. Norton, 1991.

———. "A Place for Stories: Nature, History, and Narrative." *Journal of American History*, March 1992, 1347–76.

———, ed. *Uncommon Ground: Rethinking the Human Place in Nature*. New York: W. W. Norton, 1996.

Crowcroft, Peter. *Elton's Ecologists: A History of the Bureau of Animal Populations*. Chicago: University of Chicago Press, 1991.

Cuthbert, Robert B., and Stephen G. Hoffius. *Northern Money, Southern Land: The Lowcountry Sketches of Chlotilde R. Martin*. Columbia: University of South Carolina Press, 2009.

Daniel, Pete. *Breaking the Land: The Transformation of Cotton, Tobacco, and Rice Cultures since 1880*. Urbana: University of Illinois Press, 1986.

———. "A Rogue Bureaucracy: The USDA Fire Ant Campaign of the 1950s." *Agricultural History* 64 (Spring 1990): 99–121.

Davis, Donald E. *Southern United States: An Environmental History*. Santa Barbara, Calif.: ABC-CLIO, 2006.

Davis, Frederick R. *The Man Who Saved Sea Turtles: Archie Carr and the Origins of Conservation Biology*. New York: Oxford University Press, 2007.

Davis, Jack. *An Everglades Providence: Marjory Stoneman Douglas and the Environmental Century*. Athens: University of Georgia Press, 2009.

Davis, Janet. *The Circus Age: Culture and Society under the American Big Top*. Chapel Hill: University of North Carolina Press, 2002.

de Boer, Tycho. *Nature, Business, and Community in North Carolina's Green Swamp*. Gainesville: University Press of Florida, 2008.

Delcourt, Paul A., and Hazel R. Delcourt. *Long-Term Forest Dynamics of the Temperate Zone: A Case Study of Late-Quaternary Forests in Eastern North America*. New York: Springer-Verlag, 1987.

Dillon, Kimberly Ann. "Reconstructing Aiken: Resort Development in Aiken, South Carolina, 1830–1900." MA thesis, University of South Carolina, 1997.

Donahue, Brian. *The Great Meadow: Farmers and the Land in Colonial Concord*. New Haven: Yale University Press, 2004.

Dorsey, Kurkpatrick. *The Dawn of Conservation Diplomacy: U.S.-Canadian Wildlife Protection Treaties in the Progressive Era*. Seattle: University of Washington Press, 1998.

Doughty, Robin W. *Feathers, Fashions, and Bird Preservation: A Study in Natural Protection*. Berkeley: University of California Press, 1975.

Dunlap, Thomas. *DDT: Scientists, Citizens, and Public Policy*. Princeton: Princeton University Press, 1981.

———. *Saving America's Wildlife: Ecology and the American Mind, 1850–1988*. 2nd ed. Princeton: Princeton University Press, 1988.

———. "Sport Hunting and Conservation, 1880–1920." *Environmental History Review* 12, no. 1 (1988): 51–60.

———. "That Kaibab Myth." *Journal of Forest History* 32 (1988): 60–68.

Earley, Lawrence S. *Looking for Longleaf: The Fall and Rise of an American Forest*. Chapel Hill: University of North Carolina Press, 2004.

Egerton, Frank N. "The History of Ecology: Achievements and Opportunities, Part One." *Journal of the History of Biology* 16 (1983): 259–310.

———. "The History of Ecology: Part Two." *Journal of the History of Biology* 18 (1985): 103–43.

Ellison, David L. *Healing Tuberculosis in the Woods: Medicine and Science at the End of the Nineteenth Century*. Westport, Conn.: Greenwood Press, 1994.

Engel, J. Ronald. *Sacred Sands: The Struggle for Community in the Indiana Dunes*. Middletown, Conn.: Wesleyan University Press, 1983.

Etheridge, Robbie. *Creek Country: The Creek Indians and Their World*. Chapel Hill: University of North Carolina Press, 2003.

Etheridge, Robbie, and Charles Hudson, eds. *Transformation of the Southeastern Indians, 1540–1760*. Jackson: University Press of Mississippi, 2001.

Evenden, Matthew D. "The Laborers of Nature: Economic Ornithology and the Role of Birds as Agents of Biological Pest Control in North American Agriculture, ca. 1880–1930." *Forest and Conservation History* 39 (October 1995): 172–83.

Farber, Paul. *Finding Order in Nature: The Naturalist Tradition from Linnaeus to E. O. Wilson*. Baltimore: Johns Hopkins University Press, 2000.

Feldberg, Georgina D. *Disease and Class: Tuberculosis and the Shaping of Modern North American Society*. New Brunswick: Rutgers University Press, 1995.

Fiege, Mark. *Irrigated Eden: The Making of an Agricultural Landscape in the American West*. Seattle: University of Washington Press, 1999.

Fite, Gilbert C. *Cotton Fields No More: Southern Agriculture, 1865–1980*. Lexington: University of Kentucky Press, 1984.

Fitzgerald, Deborah. *Every Farm a Factory: The Industrial Ideal in American Agriculture*. New Haven: Yale University Press, 2003.

Flader, Susan. *Thinking like a Mountain: Aldo Leopold and the Evolution of an Ecological Attitude toward Deer, Wolves, and Forests*. Lincoln: University of Nebraska Press, 1974.

Flader, Susan, and Charles Steinhacker. *The Sand Country of Aldo Leopold*. San Francisco: Sierra Club, 1973.

"Focus11: Museums and the History of Science." *Isis* 96, no. 4 (2005): 559–608.

Gates, Paul W. "Federal Land Policies in the Southern Public Land States." *Agricultural History* 53, no. 1 (1979): 206–27.

———. "Federal Land Policy in the South, 1866–1888." *Journal of Southern History* 6, no. 3 (1940): 303–30.

Geertz, Clifford. *The Interpretation of Cultures*. New York: Basic Books, 1973.

———. *Local Knowledge: Further Essays in Interpretive Anthropology*. New York: Basic Books, 1983.

Geiger, Roger L. *To Advance Knowledge: The Growth of American Research Universities, 1900–1940*. New York: Oxford University Press, 1986.

Giesen, James C. *Boll Weevil Blues: Cotton, Myth, and Power in the American South*. University of Chicago Press, 2011.

———. "'The Truth about the Boll Weevil': The Nature of Planter Power in the Mississippi Delta." *Environmental History* 14, no. 4 (2009): 683–704.

Giltner, Scott E. *Hunting and Fishing in the New South: Black Labor and White Leisure after the Civil War*. Baltimore: Johns Hopkins University Press, 2008.

Glave, Diane, and Mark Stoll, eds. *To Love the Wind and the Rain: African-Americans and Environmental History*. Pittsburgh: University of Pittsburgh Press, 2006.

Golley, Frank. *A History of the Ecosystem Concept in Ecology: More than the Sum of Its Parts*. New Haven: Yale University Press, 1996.

Gosling, Francis G. *Before Freud: Neurasthenia and the American Medical Community, 1870–1910*. Urbana: University of Illinois Press, 1987.

Graham, Otis. "Again the Backward Region?: Environmental History in and of the American South." *Southern Cultures* 6 (Summer 2000): 50–72.

Gray, Lewis Cecil. *History of Agriculture in the Southern United States to 1860*. Washington, D.C.: The Carnegie Institute, 1933.

Gregg, Sara M. "Uncovering the Subsistence Economy in the Twentieth-Century South: Blue Ridge Mountain Farms." *Agricultural History* 78, no. 4 (2004): 417–37.

Hagen, Joel B. *An Entangled Bank: The Origins of Ecosystem Ecology*. New Brunswick: Rutgers University Press, 1992.

Hahn, Steven. "Hunting, Fishing, and Foraging: Common Rights and Class Relations in the Postbellum South." *Radical History Review* 26 (1982): 37–64.

———. *The Roots of Southern Populism: Yeoman Farmers and the Transformation of the Georgia Upcountry, 1850–1890*. New York: Oxford University Press, 1983.

Hall, Marcus. *Earth Repair: The Transatlantic History of Environmental Restoration*. Charlottesville: University of Virginia Press, 2005.

———, ed., *Restoration and History: The Search for a Usable Environmental Past*. New York: Routledge, 2009.

Hamburger, Susan. "On the Land for Life: Black Tenant Farmers on Tall Timbers Plantation." *Florida Historical Quarterly* 66, no. 2 (1987): 152–59.

Haraway, Donna Jeanne. *Primate Visions: Gender, Race, and Nature in the World of Modern Science*. New York: Routledge Press, 1989.

Hart, John Fraser. "Land Use Change in a Piedmont County." *Annals of the Association of American Geographers* 70, no. 4 (1980): 492–527.

———. "Loss and Abandonment of Cleared Farm Land in the Eastern United States." *Annals of the Association of American Geographers* 58, no. 3 (1968): 417–40.

Hays, Samuel. *Beauty, Health, and Permanence: Environmental Politics in the United States, 1955–1985*. Cambridge: Cambridge University Press, 1987.

———. *Conservation and the Gospel of Efficiency: The Progressive Conservation Movement, 1890–1920*. Cambridge, Mass.: Harvard University Press, 1959.

Helms, Douglas, and Susan L. Flader, eds. *The History of Soil and Water Conservation*. Washington, D.C.: Agricultural History Society, 1985.

Henderson, Henry, and David Woolner, eds. *F.D.R. and the Environment*. New York: Palgrave Macmillan, 2005.

Herman, Daniel Justin. *Hunting and the American Imagination*. Washington, D.C.: Smithsonian Institution Press, 2001.

Hersey, Mark D. *My Work Is That of Conservation: An Environmental Biography of George Washington Carver*. Athens: University of Georgia Press, 2011.

Hirt, Paul W. *A Conspiracy of Optimism: Management of the National Forest since World War Two*. Lincoln: University of Nebraska Press, 1994.

Isenberg, Andrew C. *The Destruction of the Bison: An Environmental History, 1750–1920*. Cambridge: Cambridge University Press, 2000.

Jackson, Kenneth T. *Crabgrass Frontier: The Suburbanization of the United States*. New York: Oxford University Press, 1985.

Jacoby, Karl. *Crimes against Nature: Squatters, Poachers, Thieves, and the Hidden History of American Conservation*. Berkeley: University of California Press, 2001.

Jakle, John A. *The Tourist: Travel in Twentieth-Century America*. Lincoln: University of Nebraska Press, 1985.

Jones, William Powell. *The Tribe of Black Ulysses: African-American Lumber Workers in the Jim Crow South*. Urbana: University of Illinois Press, 2005.

Judd, Richard W. *Common Lands, Common People: The Origins of Conservation in Northern New England*. Cambridge, Mass.: Harvard University Press, 1997.

Karson, Robin. "Warren H. Manning: Pragmatist in the Wild Garden." In *Nature and Ideology: Natural Garden Design in the Twentieth Century*, edited by Joachim Wolschke-Bulmahn, 113–30. Dumbarton Oaks Colloquium on the History of Landscape Architecture, vol. 18. Washington, D.C.: Dumbarton Oaks Research Library and Collection, 1997.

King, J. Crawford. "The Closing of the Southern Range: An Exploratory Study." *Journal of Southern History* 48 (February 1982): 53–70.

Kingsland, Sharon. *The Evolution of American Ecology, 1890–2000*. Baltimore: Johns Hopkins Press, 2005.

Kirby, Jack Temple. *The Countercultural South*. Mercer University Lamar Memorial Lectures, no. 38. Athens: University of Georgia Press, 1995.

———. *Mockingbird Song: Ecological Landscapes of the South*. Chapel Hill: University of North Carolina Press, 2006.

———. *Poquosin: A Study of Rural Landscape and Society*. Chapel Hill: University of North Carolina Press, 1995.

———. *Rural Worlds Lost: The American South, 1920–1960*. Baton Rouge: Louisiana State University Press, 1987.

Koeniger, A. Cash. "Climate and Southern Distinctiveness." *Journal of Southern History* 54 (February, 1988): 21–44.

Kohler, Robert E. *All Creatures: Naturalists, Collectors, and Biodiversity, 1850–1950*. Princeton: Princeton University Press, 2006.

———. *Landscapes and Labscapes: Exploring the Lab-Field Border in Biology*. Chicago: University of Chicago Press, 2002.

———. "Paul Errington, Aldo Leopold, and Wildlife Ecology: Residential Science." *Historical Studies in the Natural Sciences* 41, no. 2 (2011): 216–54.

———. "The Ph.D. Machine: Building on the Collegiate Base." *Isis* 81, no. 4 (1990): 638–62

Kohlstedt, Sally Gregory. "Henry A. Ward: The Merchant Naturalist and American Museum Development." *Journal of the Society for the Bibliography of Natural History* 9, no. 4 (1980): 647–61.

———. "Nature, Not Books: Scientists and the Origins of the Nature-Study Movement in the 1890s." *Isis* 96 (2005): 324–52;

———. "The Nineteenth-Century Amateur Tradition: The Case of the Boston Society of Natural History." In *Science and Its Public*, edited by Gerald Holton and W. A. Blanpied. 173–90. Dordrecht: D. Reidel Publishing Company, 1976.

———. *Teaching Children Science: Hands-On Nature Study in North America, 1890–1930*. Chicago: University of Chicago Press, 2010.

Krech, Sheperd, III. *The Ecological Indian: Myth and History*. New York: W. W. Norton, 1999.

———. *Spirits of the Air: Birds and American Indians in the South*. Athens: University of Georgia Press, 2009.

Lange, Kenneth I. *A County Called Sauk: A Human History of Sauk County, Wisconsin*. Baraboo, Wis.: Sauk County Historical Society, 1976.

Langston, Nancy. *Forest Dreams, Forest Nightmares: The Paradox of Old Growth in the Inland West*. Seattle: University of Washington Press, 1995.

———. *Where Land and Water Meet: A Western Landscape Transformed*. Seattle: University of Washington Press, 2003.

Lanza, Michael L. *Agrarianism and Reconstruction Politics: The Southern Homestead Act*. Baton Rouge: Louisiana State University Press, 1990.

Lears, T. J. Jackson. *No Place of Grace: Antimodernism and the Transformation of American Culture, 1880–1920*. New York: Pantheon Books, 1981.

Lee, Kai N. *Compass and Gyroscope: Integrating Science and Politics for the Environment*. Washington, D.C.: Island Press, 1993.

Lester, Connie L., ed., *Revolution in the Land: Southern Agriculture in the 20th Century*. 18th Annual Mississippi State University History Forum. Starkville: Mississippi State University, 2002.

Livingstone, David. *Putting Science in Its Place: Geographies of Scientific Knowledge*. Chicago: University of Chicago Press, 2003.

Lutz, Tom. *American Nervousness, 1903: An Anecdotal History*. Ithaca: Cornell University Press, 1991.

Maher, Neil. "'Crazy Quilt Farming on Round Land': The Great Depression, The Soil Conservation Service, and the Politics of Landscape Change on the Great Plains During the New Deal Era." *Western Historical Quarterly* 31 (Autumn 2000): 319–39.

———. *Nature's New Deal: The Civilian Conservation Corps and the Roots of the American Environmental Movement*. New York: Oxford University Press, 2007.

Manganiello, Christopher John. "Dam Crazy with Wild Consequences: Artificial Lakes and Natural Rivers in the American South, 1845–1990." PhD diss., University of Georgia, 2010.

———. "From a Howling Wilderness to a Howling Safari: Science, Policy, Red Wolves in the American South." *Journal of the History of Biology* 42, no. 2 (May 2009): 325–59.

Marks, Stuart A. *Southern Hunting in Black and White: Nature, History, and Ritual in a Carolina Community*. Princeton: Princeton University Press, 1991.

Marshall, Lonnie A. "The Present Function of Vocational Agriculture, Schools and Other Agencies in Rural Improvement of Negroes in Leon County, Florida." MA thesis, Iowa State College, 1930.

Matthiessen, Peter. *Wildlife in America*. New York: Viking Press, 1959.

McCally, David. *The Everglades: An Environmental History*. Gainesville: University Press of Florida, 1999.

McDonald, Forrest, and Grady McWhiney. "The South from Self-Sufficiency to Peonage: An Interpretation." *American Historical Review* 85 (December 1980): 1095–118.

Mealor, W. Theodore, Jr., and Merle Prunty. "Open-Range Ranching in Southern Florida." *Annals of the Association of American Geographers* 66 (September 1976): 360–76.

Means, Bruce. “Longleaf Pine: Going, Going . . .” In *Eastern Old-Growth Forest: Prospects for Rediscovery and Recovery*, edited by Mary Byrd Davis, 210–29. Washington, D.C.: Island Press, 1996.

Meine, Curt. *Aldo Leopold: His Life and Work*. Madison: University of Wisconsin Press, 1988.

———. *Correction Lines: Essays on Land, Leopold, and Conservation*. Washington, D.C.: Island Press, 2004.

———. “The Farmer as Conservationist: Leopold on Agriculture.” In *Aldo Leopold: The Man and His Legacy*, edited by Thomas Tanner, 39–52. Ankeny, Iowa: Soil Conservation Society of America, 1987.

Meine, Curt, Michael Soulé, and Reed F. Noss. “‘A Mission-Driven Discipline’: The Growth of Conservation Biology.” *Conservation Biology* 20, no. 3 (2006): 631–51.

Mighetto, Lisa. *Wild Animals and American Environmental Ethics*. Tucson: University of Arizona Press, 1991.

Mitman, Gregg. *Breathing Space: How Allergies Shape Our Lives and Landscapes*. New Haven: Yale University Press, 2007.

———. “Hay Fever Holiday: Health, Leisure, and Place in Gilded-Age America.” *Bulletin of the History of Medicine* 77 (2003): 600–635.

———. “In Search of Health: Landscape and Disease in American Environmental History.” *Environmental History* 10, no. 2 (2005): 184–210.

———. *Reel Nature: American's Romance with Wildlife on Film*. Cambridge, Mass.: Harvard University Press, 1999.

———. *The State of Nature: Ecology, Community, and American Social Thought, 1900–1950*. Chicago: University of Chicago Press, 1992.

———. “When Nature Is the Zoo: Vision and Power in the Art and Science of Natural History.” *Osiris* 11 (1996): 117–43.

Mitman, Gregg, Michelle Murphy, and Christopher Sellers, eds. *Osiris* 19 (2004).

Morris, Christopher. “A More Southern Environmental History.” *Journal of Southern History* 75, no. 3 (2009): 581–98.

Nash, Linda. “Finishing Nature: Harmonizing Bodies and Environments in Late-Nineteenth Century California.” *Environmental History* 8, no. 1 (2003): 25–52.

———. *Inescapable Ecologies: A History of Environment, Disease, and Knowledge*. Berkeley: University of California Press, 2006.

Nash, Roderick. *Wilderness and the American Mind*. New Haven: Yale University Press, 1967.

Neel, Leon, with Paul S. Sutter and Albert G. Way. *The Art of Managing Longleaf: A Personal History of the Stoddard-Neel Approach*. Athens: University of Georgia Press, 2010.

Nelson, Lynn A. *Pharsalia: An Environmental Biography of a Southern Plantation, 1780–1880*. Athens: University of Georgia Press, 2007.

Nelson, Megan Kate. *Trembling Earth: A Cultural History of the Okefenokee Swamp*. Athens: University of Georgia Press, 2004.

Newton, Julianne Lutz. *Aldo Leopold's Odyssey*. Washington, D.C.: Island Press, 2006.

Numbers, Ronald L., and Todd L. Savitt, eds. *Science and Medicine in the Old South*. Baton Rouge: Louisiana State University Press, 1989.

Oden, Jack P. "Origins of the Southern Kraft Paper Industry, 1903–1930." *Mississippi Quarterly* 30, no. 4 (1977): 565–84.

O'Donovan, Susan Eva. *Becoming Free in the Cotton South*. Cambridge, Mass.: Harvard University Press, 2007.

Orr, Oliver H. *Saving American Birds: T. Gilbert Pearson and the Founding of the Audubon Movement*. Gainesville: University of Florida, 1992.

Ott, Katherine. *Fevered Lives: Tuberculosis in American Culture since 1870*. Cambridge, Mass.: Harvard University Press, 1996.

Otto, John Solomon. "Open-Range Cattle-Herding in Southern Florida." *Florida Historical Quarterly* 65 (January 1987): 317–34.

———. "Open-Range Cattle-Ranching in Venezuela and Florida." *Comparative Social Research* 9 (1986): 347–60.

Ownby, Ted. *Subduing Satan: Religion, Recreation, and Manhood in the Rural South, 1865–1920*. Chapel Hill: University of North Carolina Press, 1990.

Paisley, Clifton. *From Cotton to Quail: An Agricultural Chronicle of Leon County, Florida 1860–1967*. Gainesville: University Press of Florida, 1968.

———. *The Red Hills of Florida, 1528–1865*. Tuscaloosa: University of Alabama Press, 1989.

Pauly, Philip J. *Biologists and the Promise of American Life: From Meriwether Lewis to Alfred Kinsey*. Princeton: Princeton University Press, 2000.

Peck, Gunther. "The Nature of Labor: Fault Lines and Common Ground in Environmental and Labor History." *Environmental History* 11 (April 2006): 212–38.

Phillips, Sarah T. *This Land, This Nation: Conservation, Rural America, and the New Deal*. Cambridge: Cambridge University Press, 2007.

Pierce, Daniel. *The Great Smokies: From Natural Habitat to National Park*. Knoxville: University of Tennessee Press, 2000.

Powell, Katrina M. *The Anguish of Displacement: The Politics of Literacy in the Letters of Mountain Families in the Shenandoah National Park*. Charlottesville: University of Virginia Press, 2007.

Prewitt, Wiley Charles. "The Best of All Breathing: Hunting and Environmental Change in Mississippi, 1890–1980." MA Thesis, University of Mississippi, 1991.

Price, Jennifer. *Flight Maps: Adventures with Nature in Modern America*. New York: Basic Books, 1999.

Pritchard, James A. *Preserving Yellowstone's Natural Conditions: Science and the Perception of Nature*. Lincoln: University of Nebraska Press, 1999.

Pritchard, James A., Diane M. Debinski, Brian Olechnowski, and Ron Vannimwegen. "The Landscape of Paul Errington's Work." *Wildlife Society Bulletin* 34, no. 5 (2006): 1411–16.

Proctor, Nicolas W. *Bathed in Blood: Hunting and Mastery in the Old South*. Charlottesville: University of Virginia Press, 2002.

Pyne, Stephen J. *Fire in America: A Cultural History of Wildland and Rural Fire*. Princeton: Princeton University Press, 1982.

———. *Tending Fire: Coping with America's Wildland Fires*. Washington, D.C.: Island Press, 2004.

Rainger, Ronald, Keith R. Benson, and Jane Maienschein, eds. *The American Development of Biology*. Philadelphia: University of Pennsylvania Press, 1988.

———. *The Expansion of American Biology*. New Brunswick: Rutgers University Press, 1991.

Range, Willard. *A Century of Georgia Agriculture, 1850–1950*. Athens: University of Georgia Press, 1954.

Ransom, Roger L., and Richard Sutch. *One Kind of Freedom: The Economic Consequences of Emancipation*. Cambridge: Cambridge University Press, 1977.

Rappaport, Roy A. *Pigs for the Ancestors: Ritual in the Ecology of a New Guinea People*. New Haven: Yale University Press, 1967.

Ray, Janisse. *Ecology of a Cracker Childhood*. Minneapolis: Milkweed Press, 1999.

Reed, Germaine M. *Crusading for Chemistry: The Professional Career of Charles Holmes Herty*. Athens: University of Georgia Press, 1995.

Reich, Justin. "Re-creating the Wilderness: Shaping Narratives and Landscapes in Shenandoah National Park." *Environmental History* 6, no. 1 (2001): 95–117.

Reiger, John. *American Sportsmen and the Origins of Conservation*. 3rd ed. Corvallis: Oregon State University Press, 2001.

Robison, Jim, and Mark Andrews. *Flashbacks: The Story of Central Florida's Past*. Orlando: Orange County Historical Society, 1995.

Rogers, William Warren. *Antebellum Thomas County, 1828–1861*. Tallahassee: Florida State University Press, 1963.

———. *Foshalee: Quail Country Plantation, with an Overview of Leon County, Florida, and Thomas County, Georgia*. Tallahassee: Sentry Press, 1989.

———. *Pebble Hill: The Story of a Plantation*. Tallahassee: Sentry Press, 1979.

———. *Thomas County, 1865–1900*. Tallahassee: Florida State University Press, 1973.

———. *Transition to the Twentieth Century: Thomas County, Georgia, 1900–1920*. Tallahassee: Sentry Press, Thomasville Historical Society, 2002.

Rome, Adam. *The Bulldozer in the Countryside: Suburban Sprawl and the Rise of American Environmentalism*. Cambridge: Cambridge University Press, 2001.

Rosenzweig, Roy, and Elizabeth Blackmar. *The Park and the People: A History of Central Park*. Ithaca: Cornell University Press, 1992.

Rothman, Hal K. *The Greening of a Nation?: Environmentalism in the United State Since 1945*. Fort Worth: Harcourt Brace College Publishers, 1998.

Rothman, Sheila. *Living in the Shadow of Death: Tuberculosis and the Social Experience of Illness in America*. New York: Basic Books, 1994.

Rubanowice, Robert Joseph. *A Sense of Place in Southern Georgia: Birdsong Plantation, Farm, and Nature Center*. Tallahassee, Fla.: South Georgia Historical Consortium, 1994.

Rudolph, John L. "Turning Science to Account: Chicago and the General Science Movement in Secondary Education, 1905–1920." *Isis* 96 (2005): 353–89.

Saiku, Mikko. *This Delta, This Land: An Environmental History of the Yazoo-Mississippi Delta*. Athens: University of Georgia Press, 2005.

Schiff, Ashley. *Fire and Water: Scientific Heresy in the Forest Service*. Cambridge, Mass.: Harvard University Press, 1962.

Schmidt, Alfred. *The Concept of Nature in Marx*. London: NLB, 1971.

Schmitt, Peter J. *Back to Nature: The Arcadian Myth in Urban America*. New York: Oxford University Press, 1969.

Schneider, Daniel W. "Local Knowledge, Environmental Politics, and the Founding of Ecology in the United States: Stephen Forbes and 'The Lake as a Microcosm.'" *Isis* 91 (2000): 681–705.

Schuyler, David. *The New Urban Landscape: The Redefinition of City form in Nineteenth-Century America*. Baltimore: Johns Hopkins University Press, 1988.

Scott, James C. *Seeing like a State: How Certain Schemes to Improve the Human Condition Have Failed*. New Haven: Yale University Press, 1998.

Sears, John F. *Sacred Places: American Tourist Attractions in the Nineteenth Century*. New York: Oxford University Press, 1989.

Shapin, Steven. "Placing the View from Nowhere: Historical and Sociological Problems in the Location of Science." *Transactions of the Institute of British Geographers* 23 (1998): 5–12.

Shores, Elizabeth Findley. *On Harper's Trail: Roland McMillan Harper, Pioneering Botanist of the Southern Coastal Plain*. Athens: University of Georgia Press, 2008.

Silber, Nina. *The Romance of Reunion: Northerners and the South, 1865–1900*. Chapel Hill: University of North Carolina Press, 1993.

Silver, Timothy. *Mount Mitchell and the Black Mountains: An Environmental History of the Highest Peaks in Eastern America*. Chapel Hill: University of North Carolina Press, 2003.

———. *A New Face on the Countryside: Indians, Colonists, and Slaves in Southern Atlantic Forests, 1500–1800*. Cambridge: Cambridge University Press, 1990.

Smiley, David L. "The Quest for the Central Theme in Southern History." *South Atlantic Quarterly* 71 (Summer 1972): 307–25.

Smith, Neil. *Uneven Development: Nature, Capital, and the Production of Space*. New York: Basil Blackwell, 1984.

Spence, Mark David. *Dispossessing the Wilderness: Indian Removal and the Making of the National Parks*. Oxford: Oxford University Press, 1999.

Standiford, Les. *Last Train to Paradise: Henry Flagler and the Spectacular Rise and Fall of the Railroad That Crossed an Ocean*. New York: Crown Publishers, 2002.

Star, Susan Leigh. "Craft vs. Commodity, Mess vs. Transcendence: How the Right Tool Became the Wrong One in the Case of Taxidermy and Natural History." In *The Right Tools for the Job: At Work in Twentieth-Century Life Sciences*, edited by Adele E. Clarke and Joan H. Fujimura, 257–86 Princeton: Princeton University Press, 1992.

Starnes, Richard D. *Creating the Land of the Sky: Tourism and Society in Western North Carolina*. Tuscaloosa: University of Alabama Press, 2005.

———, ed. *Southern Journeys: Tourism, History, and Culture in the Modern South*. Tuscaloosa: University of Alabama Press, 2003.

Steen, Harold K., ed. *Forest and Wildlife Science in America: A History*. Durham, N.C.: Forest History Society, 1999.

Steinberg, Ted. *Down to Earth: Nature's Role in American History*. New York: Oxford University Press, 2002.

Stewart, Mart A. "If John Muir Had Been an Agrarian: American Environmental History West and South." *Environment and History* 11 (2005): 139–62.

———. "'Let Us Begin With The Weather?': Climate, Race, and Cultural Distinctiveness in the American South." In *Nature and Society in Historical Context*, edited by Mikulas Teich, Roy Porter, and Bo Gustafsson, 240–56. Cambridge: Cambridge University Press, 1997.

———. "Re-greening the South and Southernizing the Rest." *Journal of the Early Republic* 24 (Summer 2004): 242–51.

———. "Southern Environmental History." In *A Companion to the American South*, edited by John Boles, 409–23. Blackwell Publishers, 2002.

———. *What Nature Suffers to Groe: Life, Labor, and Landscape on the Georgia Coast, 1680–1920*. Athens: University of Georgia Press, 1996.

———. "'Whether Wast, Deodand, or Stray': Cattle, Culture, and the Environment in Early Georgia." *Agricultural History* 65, no. 3 (1991): 1–28.

Stoll, Steven. *Larding the Lean Earth: Soil and Society in Nineteenth Century America*. New York: Hill and Wang, 2003.

Strom, Claire, *Making Catfish Bait out of Government Boys: The Fight against Cattle Ticks and the Transformation of the Yeoman South*. Athens: University of Georgia Press, 2009.

Sutter, Paul S. *Driven Wild: How the Fight against Automobiles Launched the Modern Wilderness Movement*. Seattle: University of Washington Press, 2002.

———. "No More the Backward Region: Southern Environmental History Comes of Age." In *Environmental History and the American South: A Reader*, edited by Paul S. Sutter and Chris Manganiello, 1–24. Athens: University of Georgia Press, 2009.

———. "Representing the Resource." *Environmental History* 10, no. 1 (2005): 98–100.

———. "Terra Incognita: The Neglected History of Interwar Environmental Thought and Politics." *Reviews in American History* 29, no. 2 (2001): 289–97.

Tanner, Thomas, ed. *Aldo Leopold: The Man and His Legacy*. Ankeny, Iowa: Soil Conservation Society of America, 1987.

Tebeau, Charlton W. *A History of Florida*. Coral Gables: University of Miami Press, 1971.

Terres, John K. *Discovery: Great Moments in the lives of Outstanding Naturalists*. Philadelphia: J. B. Lippincott, 1961.

Thompson, E. P. *Whigs and Hunters: The Origins of the Black Act*. New York: Pantheon Books, 1975.

Thompson, Kenneth. "The Australian Fever Tree in California: Eucalypts and Malaria Prophylaxis." *Annals of the Association of American Geographers* 60 (June 1970): 230–44.

———. "Trees as a Theme in Medical Geography and Public Health." *Bulletin of the New York Academy of Medicine* 54, no. 5 (1978): 517–31.

———. "Wilderness and Health in the Nineteenth Century." *Journal of Historical Geography* 2 (1976): 145–61.

Tindall, George B. *The Emergence of the New South, 1913–1945*. Baton Rouge: Louisiana State University Press, 1967.

Tjossem, Sara Fairbank. "Preservation of Nature and Academic Respectability: Tensions in the Ecological Society of America, 1915–1979." PhD diss., Cornell University, 1994.

Tober, James A. *Who Owns the Wildlife?: The Political Economy of Conservation in Nineteenth-Century America*. Westport, Conn.: Greenwood Press, 1981.

Tobey, Ronald C. *Saving the Prairies: The Life Cycle of the Founding School of American Plant Ecology, 1895–1955*. Berkeley: University of California Press, 1981.

Trefethen, James B. *An American Crusade for Wildlife*. New York: Winchester Press, 1975.

Trimble, Stanley W. *Man-Induced Soil Erosion on the Southern Piedmont, 1700–1970*. Ankeny, Iowa: Soil Conservation Society of America, 1974.

———. "Perspectives on the History of Soil Erosion Control in the Eastern United States." *Agricultural History* 59, no. 2 (1985): 162–80.

Tyrell, Ian. *True Gardens of the Gods: Californian-Australian Environmental Reform, 1860–1930*. Berkeley: University of California Press, 1999.

Valencius, Conevery Bolten. *The Health of the Country: How American Settlers Understood Themselves and Their Land*. New York: Basic Books, 2002.

———. "Histories of Medical Geography." In *Medical Geography in Historical Perspective*, edited by Nicolaas A. Rupke, 3–28. London: Wellcome Trust Centre for the History of Medicine at UCL, 2000.

Vance, Rupert B. *Human Geography of the South: A Study in Regional Resources and Human Adequacy*. Chapel Hill: University of North Carolina Press, 1932.

Van Sant, Levi. "Representing Nature, Reordering Society: Eugene Odum, Ecosystem Ecology, and Environmental Politics." MA thesis, University of Georgia, 2009.

Veysey, Laurence R. *The Emergence of the American University*. Chicago: University of Chicago Press, 1965.

Warren, Louis S. *The Hunter's Game: Poachers and Conservationists in Twentieth Century America*. New Haven: Yale University Press, 1997.

Watts, W. A. "Vegetational History of the Eastern United States 25,000 to 10,000 Years Ago." In *Late Quaternary Environments*, edited by H. E. Wright Jr., 294–310. Minneapolis: University of Minnesota Press, 1983.

Wetherington, Mark V. *The New South Comes to Wiregrass Georgia, 1860–1910*. Knoxville: University of Tennessee Press, 1994.

———. *Plain Folks' Fight: The Civil War and Reconstruction in Piney Woods Georgia*. Chapel Hill: University of North Carolina Press, 2005.

White, Richard. "American Environmental History: The Development of a New Historical Field." *Pacific Historical Review* 54 (August 1985): 297–335.

———. "From Wilderness to Hybrid Landscapes: The Cultural Turn in Environmental History." *Historian* 66 (2004): 557–64.

———. *Land Use, Environment, and Social Change: The Shaping of Island County Washington*. Seattle: University of Washington Press, 1992.

———. *The Organic Machine*. New York: Hill and Wang, 1995.

Wiebe, Robert H. *A Search for Order, 1877–1920*. New York: Hill and Wang, 1967.

Williams, Michael. *Americans and Their Forests: A Historical Geography*. Cambridge: Cambridge University Press, 1989.

Williams, Raymond. *The Country and the City*. New York: Oxford University Press, 1973.

Wonders, Karen. *Habitat Dioramas: Illusions of Wilderness in Museums of Natural History*. Figura Nove Series 25. Uppsala: Acta Universitatis Upsaliensis, 1993.

Woodward, C. Vann. *Origins of the New South, 1877–1913*. Baton Rouge: Louisiana State University Press, 1951.

Worster, Donald. *Dust Bowl: The Southern Plains in the 1930s*. New York: Oxford University Press, 1979.

———, ed. *Ends of the Earth: Perspectives on Modern Environmental History*. Cambridge: Cambridge University Press, 1988.

———. *Nature's Economy: A History of Ecological Ideas*. 2nd ed. Cambridge: Cambridge University Press, 1994.

———. *Rivers of Empire: Water, Aridity, and the Growth of the American West*. New York: Pantheon Books, 1985.

———. *The Wealth of Nature: Environmental History and the Ecological Imagination*. New York: Oxford University Press, 1993.

Worster, Donald, Alfred Crosby, William Cronon, Richard White, Carolyn Merchant, and Stephen Pyne. "Roundtable in Environmental History." *Journal of American History*, March 1990, 1087–1147.

Wright, Gavin. *Old South, New South: Revolutions in the Southern Economy since the Civil War*. Baton Rouge: Louisiana State University Press, 1986.

Young, Christian C. "Defining the Range: The Development of Carrying Capacity in Management Practice." *Journal of the History of Biology* 31, no. 1 (1998): 61–83.

———. *In the Absence of Predators: Conservation and Controversy on the Kaibab Plateau*. Lincoln: University of Nebraska Press, 2002.

Youngs, Larry. "Lifestyle Enclaves: Winter Resorts in the South Atlantic States, 1870–1930." PhD diss., Georgia State University, 2001.

INDEX

ENVIRONMENTAL HISTORY AND THE AMERICAN SOUTH

Lynn A. Nelson, *Pharsalia: An Environmental Biography of a Southern Plantation, 1780–1880*

Jack E. Davis, *An Everglades Providence: Marjory Stoneman Douglas and the Environmental Century*

Shepard Krech III, *Spirits of the Air: Birds and American Indians in the South*

Paul S. Sutter and Christopher J. Manganiello, eds., *Environmental History and the American South: A Reader*

Claire Strom, *Making Catfish Bait out of Government Boys: The Fight against Cattle Ticks and the Transformation of the Yeoman South*

Christine Keiner, *The Oyster Question: Scientists, Watermen, and the Maryland Chesapeake Bay since 1880*

Mark D. Hersey, *My Work Is That of Conservation: An Environmental Biography of George Washington Carver*

Kathryn Newfont, *Blue Ridge Commons: Environmental Activism and Forest History in Western North Carolina*

Albert G. Way, *Conserving Southern Longleaf: Herbert Stoddard and the Rise of Ecological Land Management*

CPSIA information can be obtained at www.ICGtesting.com
Printed in the USA
LVOW11s1225310814

401552LV00006B/104/P